Omnidirectional Slots Antenna

Junping Geng

Omnidirectional Slots Antenna

上海交通大學出版社
SHANGHAI JIAO TONG UNIVERSITY PRESS

Springer

Junping Geng
Department of Electrical Engineering
Shanghai Jiao Tong University
Shanghai, China

ISBN 978-981-15-9091-7 ISBN 978-981-15-9089-4 (eBook)
https://doi.org/10.1007/978-981-15-9089-4

Jointly published with Shanghai Jiao Tong University Press
The print edition is not for sale in China (Mainland). Customers from China (Mainland) please order the print book from: Shanghai Jiao Tong University Press.

This Springer imprint is published by the registered company Springer Nature Singapore Pte Ltd.
The registered company address is: 152 Beach Road, #21-01/04 Gateway East, Singapore 189721, Singapore

Preface

The omnidirectional slots antenna has been studied for several years by our group. The whip antenna is an economic antenna for omnidirectional radiation on the base station or moving vehicles, but its length and gain contracted each other, and only just vertical polarization. In the real case, circular polarization omnidirectional antenna is very necessary to mobile communication in the earth, but it is difficult to realize to take place of the whip.

According to our many years antenna experiences, combining characteristics of the whip antenna, slot antenna and running TEM wave in coaxial wire, we propose the orthogonal slots array on the coaxial wire to leak the TEM wave to radiate omnidirectional circular polarization wave around the coaxial wire. After first simulation, it confirms our idea that the left hand rotated electric field appeared around the coaxial wire with several round orthogonal slots array, and there is the CP omnidirectional pattern at some frequency point. In further, we study the detail principles of the wave running in coaxial wire, the slots array combining beam in the horizontal plane, and the main beam being omnidirectional in the φ plane. And then we designed the omnidirectional RHCP slots antenna with end. Secondly, to improve gain of the omnidirectional CP antenna, we increase the rounds number of the slots array.

Thirdly, we found that the TEM wave in the coaxial wire running upward radiates RHCP wave from the orthogonal slots array. On the contrary, the TEM wave in the coaxial wire running downward radiates LHCP wave from the same orthogonal slots array too. So, we designed the dual CP omnidirectional slots antenna with down port and top port. To overcome the problem that real feeding wire to the top port affects the omnidirectional properties of the antenna, we propose the "T" shape feeding structure with two ports to feed the omnidirectional slots antenna. Both ports are in downside with enough isolation to feed RHCP and LHCP, respectively. Later, based on the T-shape feed structure, one port feeds the RHCP radiation, another port connects the inner thin coaxial wire and run through the omnidirectional slot antenna to feed a top directional UWB antenna, and they work as space diversity antenna in the half-space. Fourthly, extend the inner conductor out of the short end of the omnidirectional slot antenna, and connects a small helix, which can work as half-space cover antenna with CP performances.

Fifthly, we propose a design method of conical beam circularly polarized antenna with different beam-pointing angles. A truncated circular cone slot array antenna that radiates a Circular Polarized (CP) conical beam is presented. The direction into which the conical beam radiates depends mainly on the tilt angle of the outer conductor as well as the proper arrangement of the positions of its two slot arrays. The CP property is realized by introducing quasi-perpendicular slots that are cut into the outer conductor of a truncated circular cone. To illustrate the performance characteristics of this CP conical beam antenna, the tilt angle is set to be 45° and the beam angle is pointed at 45° from boresight within the operating band. The designed beam direction can be easily changed to meet specific application requirements. Moreover, it can be readily reconfigured to be a conformal antenna mounted on the top of a high speed mobile platform.

Finally, the reconfigurable microwave circuit with PIN diode is proposed to realize pattern diversity. We introduce the RF PIN switch into the slot, and the antenna can radiate the reconfigurable directional beam by controlling the state of PIN diodes which is lapped at the midpoint of coaxial slot. Another balance switch circuit is given to realize omnidirectional/directional switchable antenna.

Omnidirectional slots antenna really is very novel, and we found its many novel characteristics and good performances, and we still trying to tap its potential and expand its capabilities, and we will further demonstrate its magical features and application prospects.

Shanghai, China Junping Geng

Acknowledgments

The author would like to thank R. W. Ziolkowski, University of Technology Sydney, Global Big Data Technologies Centre, for several enlightening technical discussions. The author would also like to thank Prof. Ronghong Jin, University of Shanghai Jiao Tong Unversity, for multi-discussion to the contents of the book. The author would also like to thank Guanshen Chenhu, Chaofan Ren, Erwei Liu, Xuxu Cheng, Weinan Gao, Jin Zhang and Silei Yang for their support and assistance during the manuscript preparing.

Contents

About the Author

Dr. Junping Geng received his B.E degree in plastic working of metals, M.S. degree in corrosion and protection of equipment, and Ph.D. degree in circuit and system from the Northwestern Polytechnic University, Xian, China, in 1996, 1999 and 2003, respectively.

From 2003 to 2005, he was a Postdoctoral Researcher with Shanghai Jiao Tong University, Shanghai, China. In April 2005, he joined the faculty of the Electronic Engineering Department, Shanghai Jiao Tong University. He was promoted to an associated professor in 2008.

From 2010 to 2011, he was a visiting scholar with the Institute Electrical and Computer Engineering, University of Arizona, AZ, USA.

Besides, he is among the editorial board of *International Journal of Antenna and Propagation and International Journal of Aerospace Engineering*. He is also a member of Antenna Society of Chinese Institute of Electronics. And he has been the Senior member of IEEE in 2017.

He is mainly engaged in the teaching and research work on the fields covering antenna and electromagnetic theory, array signal processing, wireless communications, nanotechnology, etc.

In 2015, he received the best paper award in IEEE MAPE 2015. In 2013, he was awarded the third rank of Shanghai Award for Natural Sciences. In 2010, he received the best paper award in IEEE iWAT2010. In 2008, he won the second prize of Chinese National Technology Innovation Awards as the third co-author. In 2007, he won the first prize of Technology Innovation

Awards of the Ministry of Education as the fourth co-author. In 2009, he won the second Prize of "Hengshan Liangci" Excellent Paper.

He won the Shanghai Jiao Tong University excellent annual assessment in 2007, 2014, 2016, 2017, and 2018, respectively, and he also won the third prize for excellent teachers in 2008, excellent in the appointment period 2014–2016 in Shanghai Jiaotong University too. He won the third prize of postdoctoral award fund in 2006.

So far he had published over 350 papers at home and abroad, about 127 of them in International Journals. He has applied over 120 invention patents, and 71 of them have been granted.

Also, he has published a textbook and three monographs. He has served as an associate editor for International Journal of Antennas and Propagations, member of Technology Program Committee (TPC) and session chair for more than 15 international conferences.

He has been in charge of or involved in over 30 projects including Chinese National Fund, the State Key Program of National Natural Science, 973, Innovation group and Shanghai Research Projects.

Abbreviations

AR	Axial ratio
CP	Circularly polarized
CTL	Coaxial transmission line
DCPOA	Dual CP omnidirectional antenna
DPESA	Directional, planar equiangular spiral antenna
DRAs	Dielectric resonator antennas
FDTD	Finite difference time domain
FEM	Finite element
FIT	Finite integration technique
FI	Finite-integration
HP	Horizontal polarization
ITCC	Internal thin coaxial cable
LHCP	Left hand circularly polarized
OCPA	Omnidirectional circularly polarized antenna
OCPSA	Omnidirectional CP slot array antenna
OCPSAA	Omnidirectional CP slot array antenna
OCPTBA	Omnidirectional circularly polarized tapered beam antenna
ODCPSAA	Omnidirectional dual-CP slot array antenna
OLPA	Omnidirectional linearly polarized antenna
PDA	Polarization diversity antenna
RDRA	Rectangular dielectric resonator antenna
RHCP	Right hand circularly polarized
RLSA	Radial line slot array
SDA	Spatial diversity antenna
SP	Slant polarization
TFS	T-shaped feed structure
TL	Transmission line
UAVs	Unmanned aerial vehicles
ULA	Uniform linear array
VP	Vertical polarization
VSWR	Voltage standing wave ratio
ZPS	Zero-phase-shift

List of Figures

List of Tables

Chapter 1
Introduction

1.1 Background and Demand

The antenna is the key component of the communication, radar, navigation, radio, television and other radio system. The antenna is just like the ears and eyes of the whole communication system, which is the interface between the system and the external media. All systems based on radio wave communication are inseparable from the antenna, and for these communication systems, the selection and design of the antenna will directly affect the performance of the whole system.

Designing a proper antenna with excellent performance will greatly reduce the design pressure and cost of the whole communication system back-end, and improve the communication performance of the communication system. Since Marconi first designed the antenna to realize long-distance wireless communication in 1890s, people have done a lot of research and design on the antenna. According to different application fields and scenarios, people have designed various types of antenna.

The most outstanding characteristics of omnidirectional circularly polarized antenna are omnidirectional characteristics and original circularly polarized characteristics. Omnidirectional characteristics describe the antenna far field pattern, which means that the antenna achieves omnidirectional radiation in the horizontal plane. The most classical omnidirectional antenna is monopole antenna. Because of its simple structure, it is widely used as vehicle antenna, wireless router antenna and so on. In addition to monopole antenna, there are many ways to realize omnidirectional antenna, such as symmetrical dipole antenna, double-sided rectangular microstrip omnidirectional antenna and so on [1].

Polarization is used to describe the polarization mode of electromagnetic wave radiated by the antenna. For linear polarization, the receiving and transmitting antenna must be polarization matched to work normally [2]. For example, the transmission of TV signal usually adopts horizontal polarization mode [3]. Therefore, the position of TV receiving antenna should be adjusted to match the polarization of transmitting wave to obtain the best reception effect. The remote electromagnetic wave transmitted by AM radio is vertical polarization. Therefore, the listener adjusts

J. Geng, *Omnidirectional Slots Antenna*,
https://doi.org/10.1007/978-981-15-9089-4_1

the radio position to change the polarization state to obtain the best listening effect, that is, to achieve the polarization state matching of the transceiver.

Circular polarization refers to that the vector end track of the electromagnetic field vector radiated by the antenna is a circle. The advantages of circularly polarized antenna compared with linearly polarized antenna are mainly reflected in two aspects: first, compared with linearly polarized antenna, circularly polarized antenna can receive incoming wave of any polarization form, and its radiation wave can also be received by antenna of any polarization form; second, when circularly polarized wave is incident on symmetrical target such as plane, sphere, etc., it will produce rotation reversal, because the incoming wave and reflected wave are polarized The direction is orthogonal, so the circularly polarized antenna can suppress the reflection interference and resist the multipath effect. In many cases, the system must use circular polarization to work properly [4]. For example, the state and position of the rocket and other aircrafts are constantly changing during flight, so the polarization state of the antenna on the rocket is also constantly changing. If the linear polarization transmitting signal is used to remote control the rocket, in some cases, the linear polarization antenna on the rocket cannot receive the ground control signal, resulting in out of control. If the circular polarization antenna is used for transmitting and connecting No, that's not going to happen. In the satellite communication system, the antenna on the satellite and the antenna on the ground station work with circular polarization [5].

For the outstanding performance characteristics of omnidirectional circularly polarized antenna, it is widely used in remote sensing, detection, communication, radar, navigation, broadcasting, electronic reconnaissance and electronic detection. For example, omnidirectional circularly polarized antenna can be used to detect and interfere with electromagnetic signals in all directions except cross polarization, and it can be used for electronic reconnaissance and electronic interference in modern warfare. However, with the rapid development of electronic industry and communication technology, people's demand for communication is constantly increasing, and the requirements for modern wireless communication system in communication function, communication quality and service experience are constantly improving, which also puts forward higher and higher requirements for omni-directional antenna performance, such as requiring omni-directional circular polarization antenna to have better roundness, better axial ratio and higher gain, wider bandwidth, etc. The strict requirements bring challenges to the design of omnidirectional circularly polarized antenna, and because of this, it is of great significance and value to design an omnidirectional circularly polarized antenna that meets the requirements of the current advanced communication system and has excellent performance.

1.2 Research Development of the Omnidirectional Circularly Polarized Antennas

1.2.1 Omnidirectional Circularly Polarized Antennas

Omnidirectional circularly polarized antenna (OCPA) is ideal for wireless communication system, because it 'doesn't need strict alignment between transmitting and receiving antennas, alleviating the multipath fading as well as providing wide signal coverage. Due to these advantages, OCPAs are always concerned by researchers. The research of omnidirectional circularly polarized antenna has existed for half a century [6]. Many omnidirectional CP antennas are proposed in recent years.

Normally, there are three methods to realize omnidirectional circularly polarized antennas. The first one is to employ a circular or square antenna array which consists of multiple circularly polarized (CP) antennas to generate omnidirectional radiation pattern [7]. The CP patch antennas with cut angles are attached to the surface of a cylinder. Each CP patch antenna cover a range of angle to generate omnidirectional CP radiation pattern. The con is it requires complicate power divide networks. Based on this idea, many OCPAs were designed [8, 9].

In later researches [10], only two circularly polarized microstrip antennas with cut angles placed back-to-back can realize omnidirectional radiation pattern, as shown in [10]. Planar spiral antennas are also used as circular polarized element [11]. Rectangular loop [12], and circular loop [13] are two types of wideband circularly polarized elements used to construct wideband omnidirectional circularly polarized antenna. Helical antenna is also suitable for omnidirectional circularly polarized antenna [14]. A pair of perpendicular slots on the outer conductor of coaxial transmission line is another kind of circularly polarized element [15].

The second method is to excite two omnidirectional antennas that radiates orthogonal fields with 90° phase difference. In [16], a top-loaded monopole and four printed arc-shaped dipoles are combined. The top-loaded monopole performs vertically polarization, and four dipoles work in horizontally polarization. In [17], an UWB monopole antenna radiates vertical polarization wave, and the planar folded dipoles construct a omnidirectional antenna to radiate horizontal polarization wave. Both of them are combined into an OCPA. The antenna shown in [18] consists of four crossed dipoles, each crossed dipole element includes a horizontal dipole and a vertical dipole. A 90° phase difference between horizontally and vertically polarized component is naturally guaranteed. In [19] and [20], four tilted dipoles are combined as a combination of a loop antenna for horizontal polarization and an improved monopole antenna for vertical polarization [20]. In [21], the proposed antenna consists of 4 bended monopoles, each bended monopole is composed of a printed strip and an inverted L-shaped copper wire. The horizontal part of the inverted L-shaped copper wires produces horizontal polarization, and the vertical part of the copper wires produces vertical polarization. In [22], each antenna element comprises a dipole and a zero-phase-shift (ZPS) line loop. The in-phase fed dipole and the ZPS line loop radiate vertical and horizontal polarization omnidirectional radiation respectively. In [23], the antenna is formed by metallic strips and loops which function as electric and magnetic radiators.

Except using current loops as equivalent magnetic dipoles, in [24] slots are used as equivalent magnetic current. The antenna proposed in [25] consists of two slots for horizontally polarized radiation and planar inverted-F antennas for vertically polarized radiation.

In [26], two orthogonal circular TE_{21} modes were excited to work as antenna is proposed. There are 16 monopole sectorial loop antennas to construct the array to excite the modes. In [27], the radiating part of the antenna is a geometric structure with inclined waveguide slots, which are placed on the circular waveguide and two radially outward parallel cylindrical plates in a rotationally symmetrical manner to support circular polarization. The waves radiated from slots propagate along the radial direction in the metallic disks, in which one vertically polarized component is TEM mode, and a horizontally polarized component is TE_1 mode.

The third method is to excite an omnidirectional linearly polarized antenna (OLPA) and the parasitic or polarization conversion structures are applied to obtain the orthogonal linearly polarized component. In [28], the antenna consists of a vertical dipole and three pairs of sloped parasitic units. Similarly, in [29], the antenna has a vertical monopole and declined parasitic metal strips. In [30–34], the vertically polarized component was excited by the top-loaded monopoles, and the horizontally polarized component was excited by the arc-shaped parasitic branches. In [35], an antenna working at 28 GHz, which consists of a monopole and four radial lines printing on a single piece of substrate. In [36] the V-shaped slots in the patch and ground plane excite the horizontal polarization wave, and the top-loaded monopole produced vertical polarization both components combined. When the patch and ground is connected by vias, TM_{01} and TM_{02} modes can be excited and radiate vertical polarization wave. In [37, 38] the circular patch with curved slots radiate horizontal polarization wave, in [39] slots on the patch radiate horizontal polarization wave. In [40–42] curved branches radiate horizontal polarization wave.

In [43–46] zeroth-order resonator antenna produces omnidirectional vertical polarization, and curved branches transform half of vertical polarization into horizontal polarization. In [47], four L-shaped slots are cut on the patch of a zeroth-order resonator antenna to change the current path and get horizontally polarized wave. In [48], zeroth-order and first-order are excited at different frequencies to get left-hand circular polarization and right-hand circular polarization. In [49] zeroth-order and first-order are excited at different frequencies as well. And in [50], the antenna is based on first-order resonance, and arc arms produce horizontal polarization.

Another way is feeding the omnidirectional antenna with CP wave or adding some parasite components to the omnidirectional antenna to radiates omnidirectional CP pattern [4]. Sleeve dipole, an omnidirectional antenna, is placed in the center of outer parasite components which transfer the polarization.

There are also other ways to design omnidirectional CP antenna, like multi-layers patch antenna [51], etc. All of these methods have to compromise with some parameters (bandwidth, profile, degree of circus…), which need to considerate in real cases to gain the best antenna to utilize.

In 1955, K. S. Kelleher et al. Proposed a design method of omnidirectional circularly polarized antenna [6]. By adding a medium between the vertical polarized feed and the conical transmission surface, the vertical polarized wave radiated by omnidirectional radiation is transformed into circularly polarized wave. In 1993, the omnidirectional circular polarization antenna proposed by K. Sakaguchi et al. of Japan still drew on the idea of K.S. Kelleher et al. To produce omnidirectional circular polarization radiation electromagnetic wave by adding multiple groups of helical parasitic oscillator units outside the vertically polarized dipole antenna [52]. Although the method of on-line polarization with parasitic elements can produce omnidirectional circular polarization wave, the existence of parasitic layer increases the loss of the antenna, so the efficiency of the antenna is not high, and the circular polarization performance of the antenna is not good. Because the performance of the omnidirectional circular polarization antenna designed by changing the polarization through the parasitic structure needs to be improved, in 1995, K. Sakaguchi, based on his previous articles, fed the circular polarization spiral parasitic unit directly, making the outer ring become a spiral dipole antenna structure, realizing omnidirectional circular polarization radiation [53].

With the application of 5G technology, researchers pay more attention on mm-wave and sub-6 GHz omnidirectional circularly polarized antennas. In [54] the mm-wave omnidirectional CP antenna operates at 28 GHz. The structure is not complex so it could be fabricated cheaply. It uses square substrate to realize omnidirectional CP radiation pattern to have a better compatibility in 5G devices.

In recent years, due to the progress of technology, the performance, size and other requirements of omnidirectional circular polarization antenna are becoming higher and higher, and the research results of omnidirectional circular polarization antenna are also diverse. In 2011, B. C. Park et al. Designed an omnidirectional circularly polarized antenna [55] by using the zero-order resonant mode of the negative permittivity transmission line. The way to realize the omnidirectional circularly polarized antenna can be seen as a combination of four circularly polarized antenna elements with directional radiation, each of which covers the azimuthal angle of the antenna is overlapped by the radiation of four elements, so as to realize the omnidirectional circular polarization coverage in the horizontal radiation plane. In fact, the design idea of the omnidirectional circularly polarized antenna has been reflected in Doris I. Wu's article in 1995 [56]. Because of the negative permittivity material used, the bandwidth of the antenna is very narrow, only about 0.5%. Such a bandwidth has little significance for the high-speed communication demand of today's communication system based on Mbps. In 2012, B. C. Park improved his antenna to only have three arms. At the same time, the zero order resonance and the first order resonance of the negative permittivity transmission line were used to realize the dual band omnidirectional circular polarization antenna [57]. The bandwidth of the antenna is still very narrow, 0.5% in the low frequency and 1.1% in the high frequency. According to the final test results, the gain of the antenna is also very low, about 0–1 dbic.

In 2012, Yufeng Yu et al. Designed and completed an omnidirectional circularly polarized antenna based on the same circularly polarized principle [58], but its structure changed from a two-dimensional planar microstrip structure to a three-dimensional structure. Compared with the antenna designed by B. C. Park, Yufeng et al. The size of the antenna designed by Yu is larger, but the performance of the antenna is better. The bandwidth reaches 3.56%, and the gain is about 1.5 dB.

In 2013, A. Narbudowicz et al. Realized omni-directional circularly polarized antenna with back-to-back microstrip patch structure fed by coplanar waveguide [59], which was based on Hisao in 1998 Iwasaki's idea of designing omni-directional antenna [1], using two patch circularly polarized antenna elements to form an omni-directional circularly polarized antenna, the omni-directional property of the antenna designed in this way will be restricted to some extent, and it can be seen from the results that the omni-directional property of the antenna is not very good. In 2013, Bo Li et al. Designed an omnidirectional circularly polarized antenna [60] Based on the principle of radiating the omnidirectional circularly polarized wave by the combination of short dipole antenna and small ring antenna. In the same year, on the basis of the previously designed antenna, Bo Li designed a Polarization Reconfigurable omnidirectional circular polarization antenna [61] with the same omnidirectional circular polarization principle by adding diodes, capacitors, resistors and other lumped devices. The circularly polarized antenna obtained by this design method has a relative bandwidth of 14%, which is wider than the previous design bandwidth of those circularly polarized antennas, but the problem is that the gain of the antenna is still not high, the gain of the single circularly polarized antenna is about 0.8 dbic, and the gain of the reconfigurable circularly polarized antenna is only about 0.1 dbic.

In addition to the implementation methods of these kinds of omnidirectional circularly polarized antennas described above, K. W. Leung's team in Hong Kong has designed omnidirectional circularly polarized antennas [62, 63] in recent years using dielectric resonators. Another compact omnidirectional circularly polarized antenna has very small size, but the gain of the antennas is not high, below 2dBic, and the relative bandwidth is about 7%.

To realize high gain omnidirectional radiation, combination of folded dipoles are proposed in [64] The geometry structure of the antenna is shown in Fig. 1.1, which includes four elements ($1'$, $2'$, $3'$ and $4'$), two 180°-phase-shiftors, a feeding port and a ground plane. All elements are arranged in a " + " shape above the ground plane. Everyone likes a bow that consists of four quarter wavelength conductor poles. "AB", "BC", "CD" and "DE". "AB" and "CD" are vertical to ground plane, "BC" and "DE" are parallel to the ground plane. Assume the wavelength is λ, $|AB| = |BC| = |CD| = |DE| = |ab| = |bc| = |cd| = |de| = \lambda/4$, as in Fig. 1.1b. The whole shell shape is cylindrical, its radius is 0.625λ, height 0.25λ.The 180°-phase-shiftor can be realized by 0.5λ "U" shape pipe. The width of the arm is 0.01λ. And the manufactured antenna is shown in Fig. 1.1c.

Element $1'$ and $3'$ construct the 1st dipole, and there is an 180°-phase-shiftor between them. Element $2'$ and $4'$ constructed the 2nd dipole, and there is an 180°-phase-shiftor too. Both of these dipoles are fed by one port in parallel. The ground plane just like a mirror to produce the mirror current if there is current on the antenna.

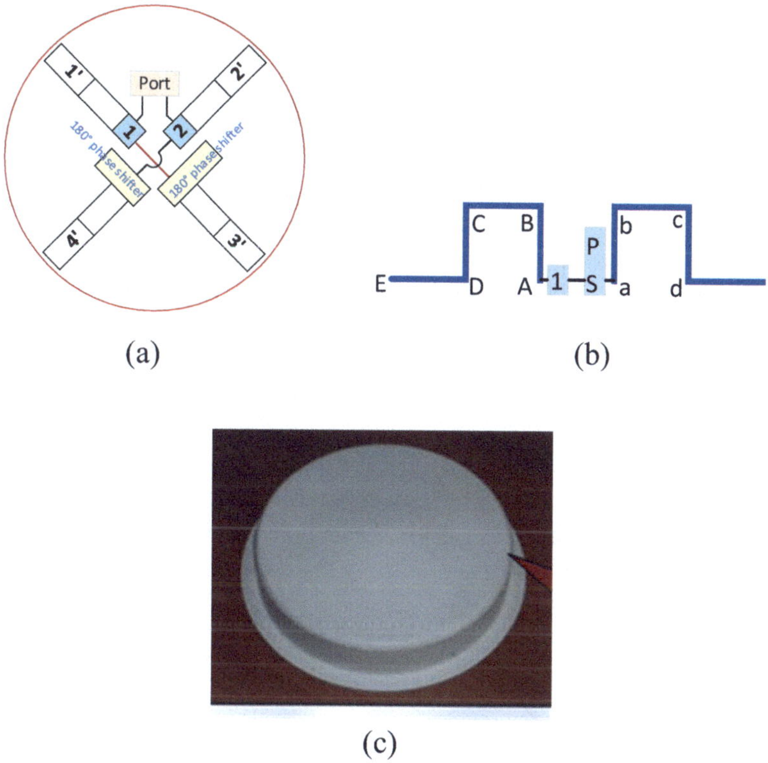

Fig. 1.1 Structure of the antenna array, 1-feeding point for the 1st dipole, 2-feeding point for the 2nd dipole: **a** Top view; **b** Side view, **c** real antenna

When the 1st dipole is fed, the current directions on "AB", "CD" are same as that on "ab", "cd" respectively, as shown in Fig. 1.2a. Meanwhile, the ground plane will introduce the mirror current too. The current magnitude distribution is shown in Fig. 1.2b, and they are similar on every part. Figure 1.2 means the 8 vertical polarized whip antennas radiate with same phase. It is similar when the 2nd dipole is excited. So, the total gain will raise to 11 dB theoretically, which is about 8 times to the half wavelength dipole.

Here, the vertical current is only on the vertical part ("AB", "CD", "ab", "cd") of the antenna, which is the key point for the vertical polarization wave.

The horizontal part ("DE", "de", "BC", "bc") of the antenna is very useful too, which can adjust the current distribution, and the detail will be analyzed in the following. Secondly, it may be a compensation to receive horizontal wave by "BC" and "bc" in some depolarization environment.

Furthermore, "DE" is very near to the ground, and they can form a capacitance C, which is very convenient to adjust the antenna impedance (to reduce the reactance of the antenna terminal) by changing the gap slightly. "de" is similar too.

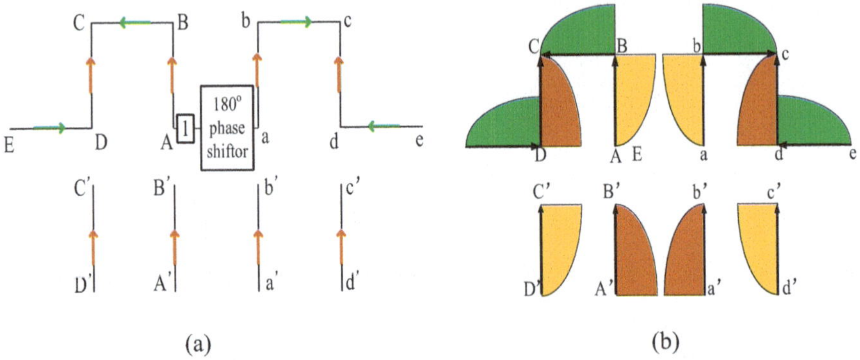

Fig. 1.2 Current distribution on the antenna, 1-feed point: **a** Current direction; **b** Current magnitude

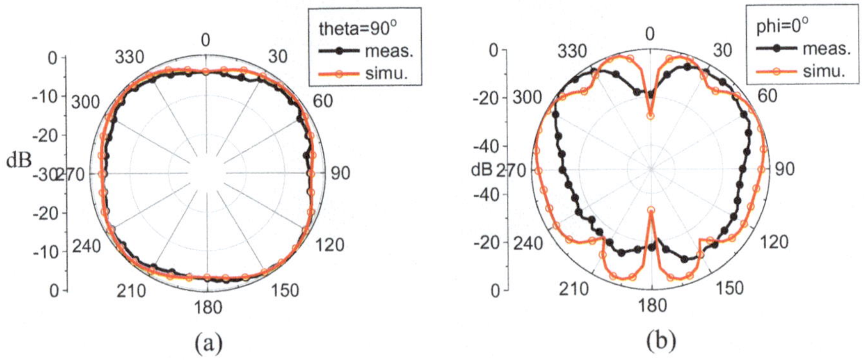

Fig. 1.3 Comparison between the simulated pattern and the measured pattern of the experiment antenna at $f = 794$ MHz: **a** theta = 90°; **b** phi = 0°.

The simulated and the measured patterns of the model 2 at $f = 794$ MHz are shown in Fig. 1.3, both of them are very close, and show good omnidirectional radiation character in horizontal plane. In Fig. 1.3b, the error is great in theta plane (phi = 0°), the main reason is from the open space and the wind

1.2.2 Research on Omnidirectional Circularly Polarized Tapered Beam Antenna

In the real, the omnidirectional circularly polarized tapered beam antenna (OCPTBA) is very useful in vehicle, remote sensing, navigation and other applications, there are many researchers pay more attention to the design of this kind of antenna. One kind of OCPTBA is planar patch antenna, which has the advantage of low profile, but it often needs more complex feed network to realize the excitation of high order resonance mode. In 1984, John Huang summarized the realization method of circularly polarized tapered beam with various angles on microstrip antenna [65]. The feed angle and phase requirements required to excite different resonant modes to achieve a circularly polarized tapered beam at different angles. A cross-section of a circularly polarized tapered beam pattern calculated theoretically. It can be seen that as the mode increases, the tapered beam gradually deviates from the normal direction of the antenna, and due to the limitation of the antenna structure, the gain decreases gradually and the antenna efficiency decreases. Based on this basic principle, there are many researches and improvements on this kind of antenna. In [66], a circularly polarized tapered beam antenna for vehicle satellite communication also is realized by exciting TM21 mode, with the tapered beam pointing at about 46 degrees. As shown in Fig. 1.4a, the antenna is fed through the bridge coupling, and the phase difference at the two feeding points can be changed by exciting different ports of the bridge, so as to realize double circular polarization. In addition, the gain and bandwidth of the antenna are improved by adding a parasitic patch with a cross slot on the top of the antenna. In order to obtain a wider bandwidth for practical use, K.L.Lau and K. M. Luk achieve a circularly polarized tapered beam about 33 degrees away from the normal by using L-Probe and aperture coupling [67]. The −10 dB impedance bandwidth is 28:2%, the 3 dB axial ratio bandwidth is 32.4%, and the maximum gain is 5:5 dbic.

Two linearly polarized antennas with orthogonal feed phase can also realize circular polarized tapered beam. In reference [68], four L-shaped patches are used to form a ring to realize horizontal polarization, and a top loaded monopole is used to realize vertical polarization, and a circularly polarized tapered beam is generated. The antenna uses pin tube to realize short circuit to ground at two different positions on L-shaped patch, so as to realize left and right circular polarization characteristics at different frequencies. But obviously, due to the radiation mechanism and the insertion loss caused by pin tube, the actual maximum gain is less than 3.2 dB, and

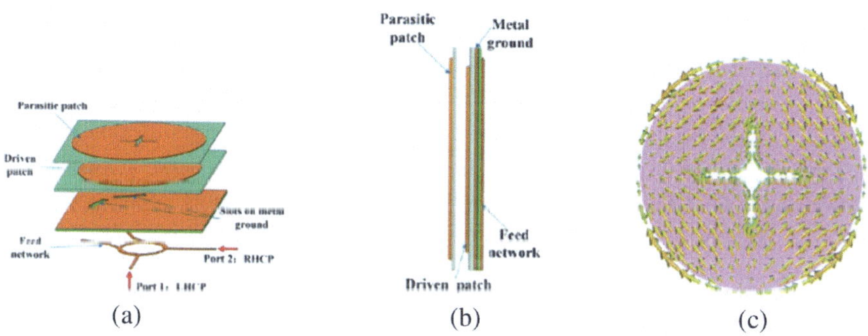

Fig. 1.4 Antenna geometry and the current distribution of the parasitic patch operated at second-order mode. **a** layered sketch, **b** side view and **c** current distribution on the top parasitic patch. Figure reproduced with permission from: [66], © 2015 IEEE

the beam direction is about 35 degrees. The principle of the antenna proposed in reference [69] is similar to that in reference [68], the broadband characteristics are realized by adding grounded through holes to the monopole to excite TM01 and TM02 modes. The axial ratio bandwidth is 14.4%, and the beam direction is about 32° away from the normal direction of the antenna. In reference [70], the circularly polarized tapered beam is realized by simultaneously exciting TM01 and TE02 modes in circular waveguide, but the 3 dB axial ratio bandwidth is only 5%, and the beam direction is about 28° away from the normal direction.

According to the development status of the above circularly polarized tapered beam antenna, each antenna proposed in the literature has its own characteristics and advantages. The main indicators of concern include section height, −10 dB return loss bandwidth, 3 dB axial ratio bandwidth, out of roundness, polarizable reconfiguration and so on, and different literatures have carried out in-depth optimization for one or several of them, and obtained the ultimate performance in some indicators. However, due to the structural limitations of the above antenna, the beam direction is relatively close to the normal direction of the antenna, so the tapered beam deviated from the normal direction cannot be realized, or the antenna efficiency decreases too much and the gain is very low when deviated from the normal direction, that is to say, the beam direction range of the conical beam circular polarization antenna is limited.

In reference [71], a radial line slot antenna is proposed. The author carves a vertical slot on the top surface of the circular waveguide and excites the slot by rotating the central probe. The equivalent magnetic flow of the slot pair along the central symmetry cancels each other in the normal direction, so the antenna pattern achieves zero depression at the top of the antenna, thus forming a circularly polarized tapered beam. The direction of the beam can be changed by changing the distance between the slot pair and the feeding center of the circular waveguide. It can be seen that, due to the limitation of the plane structure of the antenna itself, the larger the deviation angle of the required tapered beam from the normal direction is, the lower the gain of the tapered beam is. Another way is to carve mutually perpendicular gaps along the axis direction of the coaxial line [72], four circles in total, each circle is about one medium wavelength apart, equivalent to a gap array to achieve a high gain directional circular polarization beam 90° away from the normal direction. In the later, the advantages of antennas in references [71] and [72] will be discussed, and improvements will be made according to their characteristics and defects.

As for circularly polarized antenna with conical pattern, in [73] different feeding methods to excite circular microstrip antenna at higher modes are presented to generate circularly polarized conical patterns. In [74] the antenna consists of four-pair strip dipoles and slots arranged radially fed by a microstrip feed network. In [75] it is shown that a torus knot antenna can generate circularly polarized conical beam pattern. In [76] a helically slotted waveguide antenna is proposed to generate omnidirectional conical radiation.

Similar to omnidirectional circularly polarized antennas, conical patterns can be generated by exciting two orthogonal radiating fields with 90-degree phase difference. In [77, 78], an annular ring microstrip antenna is excited to radiate two TM_{21} modes with 90-degree phase difference. In [66, 67] a circular patch antenna is excited to produce TM_{21} modes with 90-degree phase difference to radiate circularly polarized wave. In [79], two orthogonal modes, TM_{01} and TE_{01} modes in a circular waveguide are excited, and they have a 90-degree phase difference. In [80] the CP property is realized by introducing quasi-perpendicular slots that are cut into the outer conductor of a truncated circular cone. In [81] the patch has an octagon-star shape, and can be considered as a superimposition of two square patches. By generating two orthogonal degenerated TM_{11} modes from the two superimposed square patches, omnidirectional cp radiation is achieved.

In [82], two crossed dipoles with conjugate impedances produce circular polarization and bidirectional pattern, by adding a reflector plate conical pattern is achieved. In [83, 84], the bidirectional pattern of two-arm spiral antenna is transformed into unidirectional conical beam by backing the antenna with a cavity.

Another way is exciting a linearly polarized array and transform the linear polarization into circular polarization. In [85], a linearly loop antenna with a perturbation element is changed to a circularly polarized antenna with conical beam. In [86] a monopole produces vertical polarization and four L-shaped patches produce horizontal polarization. In [87, 88] monopolar patch antenna provides vertical polarization and parasitic loops tubs provide horizontal polarization. In [89] center-fed circular microstrip patch provides the vertical polarization and 11 slant sector branches at the edge of circular patch generate horizontal polarization. In [90] each parasitic element has two annular slots on it, therefore at different frequencies the parasitic elements can act as resonator or transmission line to generate horizontal polarization, as a result the antenna has a wide bandwidth. In [91] the designed antenna consists of a monopolar patch fed by a probe in the center and two sets of arc hook branches. In [92–94] horizontal polarization is produced by slots. In [95] a monopole excites parasitic posts that have horizontal and vertical segments to generate circular polarization.

1.2.3 Omnidirectional Polarization Reconfigurable Circularly Polarized Antenna

Low profile compact antenna is always popular in industrial applications. Omnidirectional polarization reconfigurable circularly polarized antennas have attracted much research interest. In [93], the antenna is based on a low-profile omnidirectional CP antenna which consists of a vertically polarized microstrip patch antenna working in TM_{01}/TM_{02} modes and sequentially bended slots etched on the ground plane for radiating horizontally polarized electric field. The combined radiation from both the microstrip patch and the slots leads to a CP omnidirectional radiation pattern. The polarization reconfigurability is realized by introducing PIN diodes on the slots. By electronically controlling the states of the PIN diodes, the effective orientation of the slots on ground plane can be changed dynamically and the polarization of antenna can be altered between left-hand circular polarization and right-hand circular polarization. It operates at 2.4 GHz and realizes well axial ratio bandwidth. The pattern goes up a little bit which can be ignored. The idea of design omnidirectional CP antenna like that mentioned in [96] have already become one of main streams to realized omnidirectional CP antenna [97–100].

In [50], the vertical and horizontal polarization is generated by a parallel circular cavity and three dual arc arms, and polarization is controlled by the direction of the current on the arcs. In [101] a quad-polarization reconfigurable omnidirectional antenna is proposed. The quad-polarization antenna consists of a circular loop for horizontal polarization (HP) and a top loaded monopole for vertical polarization (VP). The circular loop is composed of four arc dipoles excited by a broadband feeding network. The left-handed circular polarization (LHCP) is obtained by simultaneously feeding the circular loop and the top-loaded monopole with different phases. Two $\lambda g/2$ phase shifters are introduced for the right-handed circular polarization (RHCP). In [102] four crossed-dipole elements around a cylinder are used to construct an omnidirectional antenna. By switching on/off four PIN diodes for each crossed-dipole element, three types of polarization reconfigurable omnidirectional antennas are realized, including vertical or horizontal polarization (VP/HP), $\pm 45°$ slant polarization (SP), and left-handed circular polarization or righthanded circular polarization (LHCP/RHCP). In [103] a coaxial slot array antenna with a T-shaped feed structure has dual circular polarization diversity, which will be discussed later.

Through the introduction of the design methods and results of all kinds of omnidirectional circularly polarized antennas, it can be seen that high gain and compact size omnidirectional antenna (OCPA) is needed widely. But it is difficult to find an omnidirectional circularly polarized antenna which is suitable for modern communication system in the existing design. Increasing gain and omnidirectional radiation performances, widening bandwidth, even flexible beam reconfiguration will show good future to OCPA.

1.3 Content and Construction of the Book

In this book, we have combed the urgent needs for high-performance omnidirectional circular polarization antenna (OCPA) in current radar and wireless communication systems, and the bottleneck problems in the design of high-performance OCPA. And find the incident point by combining the traveling wave transmission of coaxial line with the radiation of slot antenna to study the high-performance OCPA.

In this chapter, the research in OCPA, omnidirectional circularly polarized tapered beam antenna and omnidirectional polarization reconfigurable circularly polarized antenna were reviewed, especially the achievements, developing trend and problems. In Chap. 2, the related basic theory of OCPA with coaxial waveguide slot array are given. It includes the equations of transmission line, basic knowledge of coaxial wave guide, slot antenna and array.

In Chap. 3, the principle of the omnidirectional CP slot array antenna (OCPSAA) was given, and one OCPSAA in C band was designed and fabricated, it showed good omnidirectional radiation and CP performance. An improved high gain OCPSAA with more circles of slot was designed besides omni-direction and CP characteristics.

In Chap. 4, the principles of omnidirectional dual-CP slot array antenna (ODCPSAA) was studied. One ODCPSAA in C band was designed with two ports in two ends. The measured results show good omnidirectional CP radiation performances with good isolation between these two ports.

In Chap. 5, the improved ODCPSAA principle based on a compact T-shaped feed structure (TFS) was studied. And the TFS fed improved ODCPSAA was designed, which shows good polarization diversity with omnidirectional CP performances. In the further, a TFS enabled space diversity antenna was designed too, which was constructed with the OCPSA and a UWB directional spiral antenna by TFS.

In Chap. 6, a circular truncated cone slot antenna with circular polarized conical beam was presented. By cutting two sets of quasi-perpendicular slots into the outer conductor of a truncated circular cone, a CP conical beam can be produced with no limitation on the angle it points away from boresight.

In Chap. 7, a half-space covered antenna based on the OCPSA for air-ground communication was proposed. It was constructed by a slot array antenna with a helix on its top. High gain and good AR results is achieved in horizontal direction. And the top null of radiation pattern is compensated by drawing out the inner conductor of the antenna and making it into a helical.

In Chap. 8, a compact reconfigurable coaxial slot antenna based on coaxial cylinder structure with slot array and switch structure is proposed, which can realize directive radiation pattern that cover the entire horizontal plane by controlling the state of PIN diodes which is lapped at the midpoint of coaxial slot.

References

1. H. Iwasaki, A back-to-back rectangular-patch antenna fed by a CPW. IEEE Trans. Antennas Propag. **46**(10), 1527–1530 (1998)
2. W. Wei, D.M. Gong, B.S. Chen, *Antenna Principle* (National Defense Industry Press, 1985)
3. J. Lu, The modification of FM radio antenna in radio and television station. Radio Telev. Technol. **6**, 109–110 (2008)
4. H.M. Lu, Y.J. Zhao, M.Z. Zhu, *The Foundation of Electromagnetic Field and Wave* (Science Press, 2006)
5. J.F. Li, etc., A dual frequency and double circular polarization satellite communication antenna with high isolation. Acta Electronica Sinica (12), 2829–2832 (2009)
6. K. Kelleher, C. Morrow, Omnidirectional circularly-polarized antennas, in *1958 IRE International Convention Record. IEEE*, vol. 3 (1966), pp. 28–31
7. D.L. Wu, Omnidirectional circularly-polarized conformal microstrip array for telemetry applications, in *IEEE Antennas and Propagation Society International Symposium. 1995 Digest*, vol. 2 (Newport Beach, CA, USA, 1995, pp. 998–1001)
8. V. Jamnejad, J. Huang, H. Endler, F. Manshadi, Small omni-directional antenna development for Mars sample return mission, in *2001 IEEE Aerospace Conference Proceedings (Cat. No. 01TH8542)*, vol. 2 (Big Sky, MT, USA, 2001, pp. 2/843–2/851)

9. Y.Z. Yin, S.X. Gong, Q.Z. Liu, A cylindrical conformal omnidirectional microstrip antenna. J. Radio Scie. (1), 96–99+95 (1997)
10. H. Iwasaki, N. Chiba, Circularly polarised back-to-back microstrip antenna with an omnidirectional pattern. IEE Proc.: Microwaves, Antennas Propag. **146**(4), 277–281 (1999)
11. K. Nakayama, T. Kawano, H. Nakano, A conformal spiral array antenna radiating an omnidirectional circularly-polarized wave, in *Proceedings of the 1999 IEEE Antennas and Propagation Society International Symposium, APSURSI 1999, July 11, 1999–July 16, 1999, Orlando, FL, United states, F, 1999* [C] (Institute of Electrical and Electronics Engineers Inc)
12. X. Quan, R. Li, M.M. Tentzeris, A broadband omnidirectional circularly polarized antenna. IEEE Trans. Antennas Propag. **61**(5), 2363–2370 (2013)
13. Y. Fan, X. Liu, R.L. Li, A broadband circularly polarized omnidirectional antenna with circular open-loops, in *Proceedings of the 2016 IEEE International Conference on Computational Electromagnetics, ICCEM 2016, February 23, 2016 - February 25, 2016, Guangzhou, China, F, 2016* (Institute of Electrical and Electronics Engineers Inc)
14. K. Sakaguchi, N. Hasebe, A circularly polarized omnidirectional small helical antenna, in *Proceedings of the 1995 Ninth International Conference on Antennas and Propagation, ICAP '95 (Conf Publ No. 407)*, F 4–7 April 1995 (1995)
15. B. Zhou, J. Geng, X. Bai, et al., An Omnidirectional circularly polarized slot array antenna with high gain in a wide bandwidth. IEEE Antennas Wirel. Propag. Lett. **14**, 666–669 (2015)
16. K.-L. Wong, F.-R. Hsiao, C.-L. Tang, A low-profile omnidirectional circularly polarized antenna for WLAN access point, in *Proceedings of the IEEE Antennas and Propagation Society Symposium 2004 Digest held in Conjunction with: USNC/URSI National Radio Science Meeting*, June 20, 2004–June 25, 2004, Monterey, CA, United states, F, 2004. Institute of Electrical and Electronics Engineers Inc
17. X. Cai, K. Sarabandi, Broadband omni-directional circularly polarized antenna based on vertically and horizontally polarized elements, in *Proceedings of the 2016 IEEE Antennas and Propagation Society International Symposium, APSURSI 2016*, June 26, 2016–July 1, 2016, Fajardo, Puerto rico, F, 2016. Institute of Electrical and Electronics Engineers Inc
18. C.-Y. Liu, Q.-X. Chu, Omnidirectional circularly polarized antenna based on crossed dipoles, in *Proceedings of the 9th International Conference on Microwave and Millimeter Wave Technology, ICMMT 2016*, June 5, 2016–June 8, 2016, Beijing, China, F, 2016. Institute of Electrical and Electronics Engineers Inc
19. Quan X, Li R. Broadband dual-polarized omnidirectional antennas, in *Proceedings of the Joint 2012 IEEE International Symposium on Antennas and Propagation and USNC-URSI National Radio Science Meeting, APSURSI 2012*, July 8, 2012–July 14, 2012, Chicago, IL, United states, F, 2012. Institute of Electrical and Electronics Engineers Inc
20. Y. Fan, X. Quan, Y. Pan et al., Wideband Omnidirectional Circularly Polarized Antenna Based on Tilted Dipoles [J]. IEEE Trans. Antennas Propag. **63**(12), 5961–5966 (2015)
21. Y. Yufeng, S. Zhongxiang, H. Sailing, Compact omnidirectional antenna of circular polarization. IEEE Antennas Wirel. Propag. Lett. **11**, 1466–1469 (2012)
22. J. Shi, X. Wu, X. Qing et al., An omnidirectional circularly polarized antenna array. IEEE Trans. Antennas Propag. **64**(2), 574–581 (2016)
23. W. Lin, R.W. Ziolkowski, Compact, high directivity, omnidirectional circularly polarized antenna array. IEEE Trans. Antennas Propag. **67**(7), 4537–4547 (2019)
24. J. Liu, Y. Li, Z. Liang et al., A planar quasi-magnetic–electric circularly polarized antenna. IEEE Trans. Antennas Propag. **64**(6), 2108–2114 (2016)
25. J. Wu, K. Sarabandi, Compact omnidirectional circularly polarized antenna. IEEE Trans. Antennas Propag. **65**(4), 1550–1557 (2017)
26. B. Yektakhah, K. Sarabandi, A wideband circularly polarized omnidirectional antenna based on excitation of two orthogonal circular TE21 modes. IEEE Trans. Antennas Propag. **65**(8), 3877–3888 (2017)
27. C. Turkmen, M. Secmen, Omnidirectional and Circularly Polarized Slotted Antenna Array With Increased Bandwidth Performance by Using Nonidentical Waveguide Slots [J]. Radio Science **53**(11), 1406–1418 (2018)

28. K. Sakaguchi, N. Hasebe, A circularly polarized omnidirectional antenna; proceedings of the 1993 Eighth International Conference on Antennas and Propagation, F 30 March-2 April 1993, 1993
29. J.M. Fernández, J.L. Masa-Campos, M. Sierra-Pérez, Circularly polarized omnidirectional millimeter wave monopole with parasitic strip elements. Microwave Opt. Technol. Lett. **49**(3), 664–668 (2007)
30. B. Li, S.-W. Liao, Q. Xue, Omnidirectional circularly polarized antenna combining monopole and loop radiators. IEEE Antennas Wirel. Propag. Lett. **12**, 607–610 (2013)
31. B. Li, Q. Xue, Polarization-reconfigurable omnidirectional antenna combining dipole and loop radiators. IEEE Antennas Wirel. Propag. Lett. **12**, 1102–1105 (2013)
32. D. Wu, X. Chen, L. Yang, et al., Compact and low-profile omnidirectional circularly polarized antenna with four coupling Arcs for UAV applications. IEEE Antennas Wirel. Propag. Lett. **16**, 2919–2922 (2017)
33. L. Yang, Z.-Y. Zhang, L. Ji, et al., Wideband omnidirectional circularly polarized patch antenna with multi-resonant structure. Prog. Electromagnet. Res. C **78**, 47–55 (2017)
34. Liu C-Y, Chu Q-X. A compact omnidirectional circularly polarized antenna; proceedings of the 12th European Conference on Antennas and Propagation, EuCAP 2018, April 9, 2018 - April 13, 2018, London, United kingdom, F, 2018 [C]. Institution of Engineering and Technology.
35. W. Lin, R.W. Ziolkowski, T.C. Baum, 28 GHz compact omnidirectional circularly polarized antenna for device-to-device communications in the future 5G systems. IEEE Trans. Antennas Propag. **65**(12), 6904–6914 (2017)
36. C. Guo, R. Yang, W. Zhang, Compact omnidirectional circularly polarized antenna loaded with complementary V-shaped slits. IEEE Antennas Wirel. Propag. Lett. **17**(9), 1593–1597 (2018)
37. D. Yu, S.-X. Gong, Y.-T. Wan, et al., Omnidirectional dual-band dual circularly polarized microstrip antenna using TM01 and TM02 modes. IEEE Antennas Wirel. Propag. Lett. **13**, 1104–1107 (2014)
38. Y. Liu, X. Li, L. Yang et al., A dual-polarized dual-band antenna with omni-directional radiation patterns. IEEE Trans. Antennas Propag. **65**(8), 4259–4262 (2017)
39. Y. Shi, J. Liu, Wideband and low-profile omnidirectional circularly polarized antenna with slits and shorting-vias. IEEE Antennas Wirel. Propag. Lett. **15**, 686–689 (2016)
40. H.H. Tran, N. Nguyen-Trong, H.C. Park, A compact wideband omnidirectional circularly polarized antenna using TM01 mode with capacitive feeding. IEEE Antennas Wirel. Propag. Lett. **18**(1), 19–23 (2019)
41. Y.M. Pan, S.Y. Zheng, B.J. Hu, Wideband and low-profile omnidirectional circularly polarized patch antenna. IEEE Trans. Antennas Propag. **62**(8), 4347–4351 (2014)
42. X. Chen, W. Zhang, L. Han, et al., Wideband circularly polarized antenna realizing omni-directional radiation in the wider Azimuth planes. IEEE Antennas Wirel. Propag. Lett. **16**, 2461–2464 (2017)
43. J. An, G.-M. Wang, C.-X. Zhang et al., Compact circularly polarized omnidirectional microstrip antenna. Microwave Opt. Technol. Lett. **51**(11), 2643–2646 (2009)
44. B.-C. Park, J.-H. Lee, Omnidirectional circularly polarized antenna based on meta material transmission line, in *Proceedings of the 2009 IEEE International Symposium on Antennas and Propagation and USNC/URSI National Radio Science Meeting, APSURSI 2009*, June 1, 2009–June 5, 2009, North Charleston, SC, United states, F, 2009. Institute of Electrical and Electronics Engineers Inc
45. L. Yang, N.-W. Liu, Z.-Y. Zhang, et al., A novel single feed omnidirectional circularly polarized antenna with wide AR bandwidth. Prog. Electromagnet. Res. C **51**, 35–43 (2014)
46. J. Li, Y. Huang, G. Wen, Design of a compact tri-band omnidirectional circularly polarized antenna, in *Proceedings of the 2018 IEEE Antennas and Propagation Society International Symposium and USNC/URSI National Radio Science Meeting, APSURSI 2018*, July 8, 2018–July 13, 2018, Boston, MA, United states, F, 2018 [C]. Institute of Electrical and Electronics Engineers Inc.

47. G. Xu, Y. Sun, M. Yang, et al., An omnidirectional circularly polarized patch antenna with bandwidth broadening, in *Proceedings of the 9th International Conference on Microwave and Millimeter Wave Technology, ICMMT 2016*, June 5, 2016–June 8, 2016, Beijing, China, F, 2016. Institute of Electrical and Electronics Engineers Inc
48. B.C. Park, J.H. Lee, Dual-band omnidirectional circularly polarized antenna using zeroth- and first-order modes. IEEE Antennas Wirel. Propag. Lett. **11**, 407–410 (2012)
49. X. Chen, L. Han, X. Chen et al., Dual-band circularly polarized antenna using mu-negative transmission lines. IEEE Antennas Wirel. Propag. Lett. **17**(7), 1190–1194 (2018)
50. D. Tang, X. Kong, S. Liu, Design of a low profile omnidirectional circular polarized reconfigurable antenna for Beidou application, in *Proceedings of the 2018 IEEE International Conference on Computational Electromagnetics, ICCEM 2018*, March 26, 2018–March 28, 2018, Chengdu, China, F, 2018 [C]. Institute of Electrical and Electronics Engineers Inc
51. H. Iwasaki, N. Chiba, Circularly polarised back-to-back microstrip antenna with an omnidirectional pattern. IEE Proc. - Microwaves, Antennas Propag. **146**(4), 277–281 (1999)
52. K. Sakaguchi, N. Hasebe, A circularly polarized omnidirectional antenna, *in Eighth International Conference on Antennas and Propagation* (IET, 1993), pp. 477–480
53. Sakaguchi K, Hasebe N. A circularly polarized omnidirectional small helical antenna, in *Antennas and Propagation, 1995., Ninth International Conference on (Conf. Publ. No. 407)* (IET, 1995), pp. 492–495
54. P.F. Hu, *Research and Design of Millimeter Wave Array Antenna* (Xi'an University of Electronic Science and technology, 2018)
55. B.C. Park, J.H. Lee, Omnidirectional circularly polarized antenna utilizing zeroth-order resonance of epsilon negative transmission line. IEEE Trans. Antennas Propag. **59**(7), 2717–2721 (2011)
56. D.L. Wu, Omnidirectional circularly-polarized conformal microstrip array for telemetry applications, in *Antennas and Propagation Society International Symposium*. AP-S. Digest, vol. 2 (IEEE, 1995), pp. 998–1001
57. B.C. Park, J.H. Lee, Dual-band omnidirectional circularly polarized antenna using zeroth-and first-order modes. Antennas Wirel. Propag. Lett., IEEE **11**, 407–410 (2012)
58. Y. Yu, Z. Shen, S. He, Compact omnidirectional antenna of circular polarization. Antennas Wirel. Propag. Lett., IEEE **11**, 1466–1469 (2012)
59. A. Narbudowicz, X.L. Bao, M.J. Ammann, Dual-band omnidirectional circularly polarized antenna. IEEE Trans. Antennas Propag. **61**(1), 77–83 (2013)
60. B. Li, S.W. Liao, Q. Xue, Omnidirectional circularly polarized antenna combining monopole and loop radiators. Antennas Wirel. Propag. Lett., IEEE **12**, 607–610 (2013)
61. B. Li, Q. Xue, Polarization-reconfigurable omnidirectional antenna combining dipole and loop radiators. Antennas Wirel. Propag. Lett., IEEE **12**, 1102–1105 (2013)
62. Y.M. Pan, K.W. Leung, K. Lu, Omnidirectional linearly and circularly polarized rectangular dielectric resonator antennas. IEEE Trans. Antennas Propag. **60**(2), 751–759 (2012)
63. W.W. Li, K.W. Leung, Omnidirectional circularly polarized dielectric resonator antenna with top-loaded Alford loop for pattern diversity design. IEEE Trans. Antennas Propag. **61**(8), 4246–4256 (2013)
64. J. Geng, R. Jin, W. Wang, W. He, M. Ding, Q. Wu, X. Rui, G. Yang, Z. Fang, A new quasi-omnidirectional vertical polarisation antenna with low profile and high gain for DTV on vehicle. Microwaves, Antennas Propag., IET**1**(4), 918–924, (2007)
65. J. Huang, Circularly polarized conical patterns from circular microstrip antennas. IEEE Trans. Antennas Propag. **32**(9), 991–994 (1984)
66. X. Bai, X. Liang, M. Li et al., Dual-circularly polarized conical-beam microstrip antenna. Antennas Wirel. Propag. Lett. IEEE **14**, 482–485 (2015)
67. K.L. Lau, K.M. Luk, A wideband circularly polarized conical-beam patch antenna. IEEE Trans. Antennas Propag. **54**(5), 1591–1594 (2006)
68. J.S. Row, M.C. Chan, Reconfigurable circularly-polarized patch antenna with conical beam. IEEE Trans Antennas Propag. **58**(8), 2753–2757 (2010)

69. W. Lin, W. Hang, Circularly polarized conical-beam antenna with wide bandwidth and low profile. IEEE Trans. Antennas Propag. **62**(12), 5974–5982 (2014)
70. S. Shan Qi, W. Wu, D. Gang Fang, Singly-fed circularly polarized circular aperture antenna with conical beam. IEEE Trans. Antennas Propag. **61**(6), 3345–3349 (2013)
71. J. Takada, A. Tanisho, K. Ito et al., Circularly polarised conical beam radial line slot antenna. Electron. Lett. **30**(21), 1729–1730 (1994)
72. B. Zhou, J. Geng, X. Bai et al., An omnidirectional circularly polarized slot array antenna with high gain in a wide bandwidth. IEEE Antennas Wirel. Propag. Lett. **14**, 666–669 (2015)
73. J. Huang, Circularly polarized conical patterns from circular microstrip antennas. IEEE Trans. Antennas Propag. AP-32(9), 991–994 (1984)
74. K. Ito, J.P. Daniel, J.M. Lenormand, Printed antenna composed of strip dipoles and slots generating circularly polarized conical patterns, in *Proceedings of the International Symposium Digest—Antennas and Propagation—1989, June 26, 1989–June 30, 1989, San Jose, CA, USA, F*, 1989. Publ by IEEE
75. S.V. Kumar, A.R. Harish, Generation of circularly polarized conical beam pattern using torus knot antenna. IEEE Trans. Antennas Propag. **65**(11), 5740–5746 (2017)
76. Y. Liu, M. Li, K. Song, et al., Leaky-wave antenna with switchable omnidirectional conical radiation via polarization handedness. IEEE Trans. Antennas Propag., **1** (2019)
77. H. Ohmine, Y. Sunahara, M. Matsunaga, Annular-ring microstrip antenna fed by a co-planar feed circuit for mobile satellite communication use. IEEE Trans. Antennas Propag. **45**(6), 1001–1008 (1997)
78. X. Chen, G. Fu, S.X. Gong, et al., Single-feeding circularly polarized TM21-mode annular-ring microstrip antenna for mobile satellite communication. Progress Electromagn. Res. Lett. **20**, 147–156 (2011)
79. S.S. Qi, W. Wu, D.G. Fang, Singly-fed circularly polarized circular aperture antenna with conical beam. IEEE Trans. Antennas Propag. **61**(6), 3345–3349 (2013)
80. Chenhu G, Geng J, Zhou H, et al., Truncated circular cone slot antenna array that radiates a circularly polarized conical beam. IEEE Antennas Wirel. Propag. Lett. **16**, 2574–2577 (2017)
81. Y. Shi, J. Liu, A circularly polarized octagon-star-shaped microstrip patch antenna with conical radiation pattern. IEEE Trans. Antennas Propag. **66**(4), 2073–2078 (2018)
82. Nesic A, Brankovic V, Radnovic I. New circularly polarized planar printed antenna with conical radiation pattern, in *Proceedings of the 1998 IEEE Antennas and Propagation Society International Symposium, APSURSI 1998*, June 21, 1998–June 26, 1998, Atlanta, GA, United states, F, 1998. Institute of Electrical and Electronics Engineers Inc
83. H. Nakano, H. Oyanagi, J. Yamauchi, A wideband circularly polarized conical beam from a two-arm spiral antenna excited in phase. IEEE Trans. Antennas Propag. **59**(10), 3518–3525 (2011)
84. B. Wen, Z.-H. Yan, K. Chen, et al., A low-profile two-arm Archimedean spiral antenna radiating a circularly polarized normal beam or conical beam, in *Proceedings of the 2013 International Workshop on Microwave and Millimeter Wave Circuits and System Technology, MMWCST 2013*, October 24, 2013–October 25, 2013, Emeishan, Sichuan, China, F, 2013. IEEE Computer Society
85. H. Nakano, K. Fujimori, J. Yamauchi, Low-profile conical beam loop antenna with an electromagnetically coupled feed system. IEEE Trans. Antennas Propag. **48**(12), 1864–1866 (2000)
86. R. Jeen-Sheen, C. Ming-Che, Reconfigurable circularly-polarized patch antenna with conical beam. IEEE Trans. Antennas Propag. **58**(8), 2753–2757 (2010)
87. W. Lin, H. Wong, Circularly polarized conical-beam antenna with wide bandwidth and low profile. IEEE Trans. Antennas Propag. **62**(12), 5974–5982 (2014)
88. W. Lin, H. Wong, Polarization reconfigurable wheel-shaped antenna with conical-beam radiation pattern. IEEE Trans. Antennas Propag. **63**(2), 491–499 (2015)
89. D. Yu, Y.-T. Wan, H. B. Zhang, Low-profile wideband circularly polarized microstrip antenna with conical radiation pattern. Prog. Electromagn. Res. C **72**, 81–89 (2017)

90. H. Xu, J. Zhou, Q. Wu et al., Wideband low-profile SIW cavity-backed circularly polarized antenna with high-gain and conical-beam radiation. IEEE Trans. Antennas Propag. **66**(3), 1179–1188 (2018)
91. He W, Zhang L, He Y, et al., A wideband circularly polarized antenna with conical-beam radiation, in *Proceedings of the 2019 IEEE International Conference on RFID Technology and Applications, RFID-TA 2019, September 25, 2019–September 27, 2019, Pisa, Italy*, F, 2019. Institute of Electrical and Electronics Engineers Inc
92. D. Yu, S.-X. Gong, Y.-T. Wan et al., Wideband omnidirectional circularly polarized patch antenna based on vortex slots and shorting vias. IEEE Trans. Antennas Propag. **62**(8), 3970–3977 (2014)
93. Y.-M. Cai, S. Gao, Y. Yin et al., Compact-size low-profile wideband circularly polarized omnidirectional patch antenna with reconfigurable polarizations. IEEE Trans. Antennas Propag. **64**(5), 2016–2021 (2016)
94. H.-Y. Zhang, F.-S. Zhang, C. Wang, et al., Dual-band omnidirectional circularly polarized patch antenna with etched slots and shorting vias. Prog. Electromag. Res. C **73**, 167–176 (2017)
95. S. Karki, M. Sabbadini, K. Alkhalifeh et al., Metallic monopole parasitic antenna with circularly polarized conical patterns. IEEE Trans. Antennas Propag. **67**(8), 5243–5252 (2019)
96. Y. Cai, S. Gao, Y. Yin, W. Li, Q. Luo, Compact-size low-profile wideband circularly polarized omnidirectional patch antenna with reconfigurable polarizations. IEEE Trans. Antennas Propag. **64**(5), 2016–2021 (2016)
97. B. Park, J. Lee, Omnidirectional circularly polarized antenna utilizing zeroth-order resonance of epsilon negative transmission line. IEEE Trans. Antennas Propag. **59**(7), 2717–2721 (2011)
98. Y. Ma, J. Li, R. Xu, Design of an omnidirectional circularly polarized antenna. IEEE Antennas Wirel. Propag. Lett. **16**, 226–229 (2017)
99. B. Park, J. Lee, Dual-band omnidirectional circularly polarized antenna using zeroth- and first-order modes. IEEE Antennas Wirel. Propag. Lett. **11**, 407–410 (2012)
100. D. Yu, S. Gong, Y. Wan, W. Chen, Omnidirectional dual-band dual circularly polarized microstrip antenna using ${\rm TM}_{01}$ and ${\rm TM}_{02}$ Modes. IEEE Antennas Wirel. Propag. Lett. **13**, 1104–1107 (2014)
101. Y. Cui, C. Qi, R. Li, A low-profile broadband quad-polarization reconfigurable omnidirectional antenna. IEEE Trans. Antennas Propag. **67**(6), 4178–4183 (2019)
102. Y. Fan, R. Li, Y. Cui, Development of polarisation reconfigurable omnidirectional antennas using crossed dipoles. IET Microwaves Antennas Propag. **13**(4), 485–491 (2019)
103. J. Geng, R.W. Ziolkowski, K. Wang, et al., Dual CP polarization diversity and space diversity antennas enabled by a compact T-shaped feed structure. IEEE Access (7), 96284–96296 (2019)

Chapter 2
Basic Theory of Omnidirectional Antenna and Slot Array

2.1 The Basic Theory of the Transmission Line

2.1.1 The Uniform Transmission Line

2.1.1.1 The Equation of the Uniform Transmission Line and Its General Solution [1]

Microwave transmission line (TL) is a general term for various forms of transmission system used to transmit microwave information and energy. Its function is to guide electromagnetic wave to transmit in a certain direction, so it is also called guided wave system, and the guided electromagnetic wave is called guided wave. Generally, guided wave system with constant section size, shape, medium distribution, material and boundary conditions is called regular guided wave system, also known as uniform transmission line.

The guided wave system composed of uniform transmission lines can be equivalent to the uniform parallel double conductor system as shown in Fig. 2.1.

Assume the voltage u and current i are the equation of space and time, then

$$\begin{cases} u = u(z, t) \\ i = i(z, t) \end{cases} \tag{2.1}$$

From the Kirchhoff principle, it can be concluded

$$\begin{cases} -\dfrac{\partial u}{\partial z} = Ri + L\dfrac{\partial i}{\partial t} \\ -\dfrac{\partial i}{\partial z} = Gu + C\dfrac{\partial u}{\partial t} \end{cases} \tag{2.2}$$

J. Geng, *Omnidirectional Slots Antenna*,
https://doi.org/10.1007/978-981-15-9089-4_2

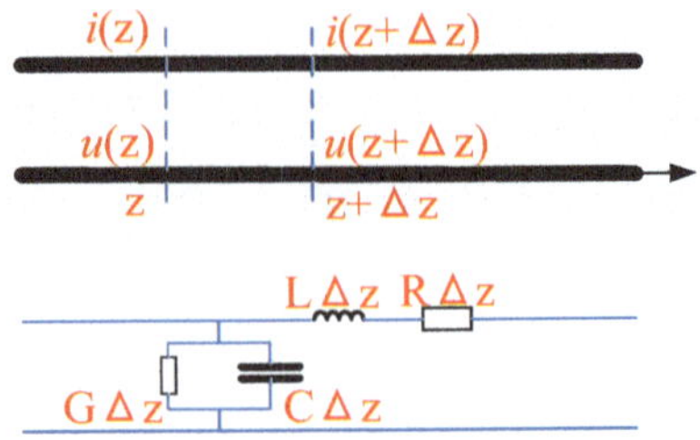

Fig. 2.1 Uniform transmission line and the equivalent circuit

When $\Delta z \to 0$,

$$\begin{cases} -u(z+\Delta z, t) + u(z, t) = \left[Ri(z, t) + L\dfrac{\partial i(z, t)}{\partial t} \right] \Delta z \\ -i(z+\Delta z, t) + i(z, t) = \left[Gu(z, t) + C\dfrac{\partial u(z, t)}{\partial t} \right] \Delta z \end{cases} \tag{2.3}$$

Equation (2.3) is just the uniform transmission line equation, and called the telegraph equation.

Here, if we just study the change of time harmonic (sine or cosine), then

$$\begin{cases} u(z, t) = R_e\left[U(z)e^{j\omega t}\right] \\ i(z, t) = R_e\left[I(z)e^{j\omega t}\right] \end{cases} \tag{2.4}$$

In (2.4), $U(z)$ and $I(z)$ are only related to z, they are the effective comprehensive voltage or current value at position z in the transmission line. It was put into Eq. (2.3), then

$$\begin{cases} -\dfrac{du}{dz} = (R + j\omega L)I = ZI \\ -\dfrac{dI}{dz} = (G + j\omega C)U = YU \end{cases} \tag{2.5}$$

In Eq. (2.5), $Z = R + j\omega L$, $Y = G + j\omega C$, they are called per-unit length series impedance and per-unit length parallel admittance of transmission line respectively.

To the lossless transmission line, $R = 0$, $G = 0$, the Eq. (2.5) can be solved,

$$\begin{cases} U(z) = u_{+}(z, t) + u_{-}(z, t) = A_1 e^{j\beta z} + A_2 e^{j\beta z} \\ I(z) = i_{+}(z, t) + i_{-}(z, t) = \dfrac{1}{z_0}(A_1 e^{-j\beta z} - A_2 e^{j\beta z}) \end{cases} \tag{2.6}$$

Fig. 2.2 Boundary conditions coordinate system

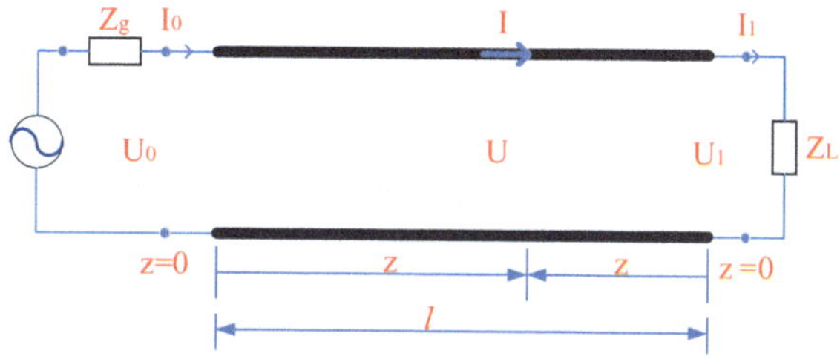

In Eq. (2.6), the phase shift constant $\beta = \omega\sqrt{LC}$, the characteristic impedance $Z_0 = \sqrt{\frac{L}{C}}$, L the unit length inductance, C is the per-unit length capacitance. Equation (2.6) is called the general solution of the transmission equation, A_1 and A_2 can be determined by the confirmed boundary conditions.

Only under the detailed boundary conditions, the general solution can be converted to the concrete solution. As shown in Fig. 2.2, two sets of coordinates systems are established, and coordinate z is from the source, coordinate z' is from the terminal.

Under the terminal condition, the Z_l and U_l are given, so coefficients can be achieved from the Eq. (2.6)

$$\begin{cases} A_1 = \dfrac{U_l + Z_0 I_l}{2} e^{j\beta l} \\ A_2 = \dfrac{U_l - Z_0 I_l}{2} e^{-j\beta l} \end{cases} \tag{2.7}$$

And then the general solutions can be written

$$\begin{cases} U(z) = \dfrac{1}{2}(U_l + Z_0 I_l)e^{j\beta(l-z)} + \dfrac{1}{2}(U_l - Z_0 I_l)e^{-j\beta(l-z)} \\ I(z) = \dfrac{1}{2Z_0}(U_l + Z_0 I_l)e^{j\beta(l-z)} - \dfrac{1}{2Z_0}(U_l - Z_0 I_l)e^{-j\beta(l-z)} \end{cases} \tag{2.7a}$$

If we use z' (from the terminal) coordinate system, $z' = l - z$. Considering the Euler equations,

$$\begin{cases} e^{j\beta z'} = \cos\beta z' + j\sin\beta z' \\ e^{-j\beta z'} = \cos\beta z' - j\sin\beta z' \end{cases} \tag{2.8}$$

Then,

$$\begin{cases} U(z') = U(l)\cos\beta z' + jZ_0 I(l)\sin\beta z' \\ I(z') = j\dfrac{U(l)}{Z_0}\sin\beta z' + I(l)\cos\beta z' \end{cases} \tag{2.9}$$

2.1.1.2 The Character Parameters of the TL

1. Characteristic impedance

The ratio of voltage to current of the traveling wave on the transmission line is defined as the characteristic impedance of the transmission line, Z_0. To the lossless coaxial wire with radius of a and b of the inner conductor and outer conductor, its characteristic impedance is

$$Z_0 = \frac{60}{\sqrt{\varepsilon_r}} \ln \frac{b}{a} \tag{2.10}$$

2. The wavelength of the transmission line

The relation between the wavelength of the transmission line λ_g and the wavelength in free space λ_0 can be written as

$$\lambda_g = \frac{\lambda_0}{\sqrt{\varepsilon_r}} \tag{2.11}$$

3. Electrical length

The electrical length of the transmission line can be expressed as

$$\theta = \beta l \tag{2.12}$$

2.1.2 The State Parameters and Impedance of the TL

2.1.2.1 Input Impedance

At any position z' in the TL, the ratio of the incident voltage and current is called the input impedance, and written as $Z_{in}(z')$,

$$Z(z') = U(z')/I(z') \tag{2.13}$$

From Eq. (2.8), it can be given

$$Z(z') = \frac{\cos \beta z' U_l + jZ_0 \sin \beta z' I_l}{j\frac{1}{2} \sin \beta z' U_l + \cos \beta z' I_l} = Z_0 \frac{Z_l + jZ_0 \tan \beta z'}{Z_0 + jZ_l \tan \beta z'} \tag{2.14}$$

Here, $Z_l = U_l/I_l$ is the load impedance, and the relation between the input impedance and the load impedance can be gotten by Eq. (2.14). The input impedance has the following properties,

(1) Z_l can be converted to $Z(z')$ by the TL with length z', so the TL can be regarded as impedance transformer.
(2) The impedance is periodical. The period of $\tan \beta z'$ is π, which means the integer times of the half wavelength, $m\lambda_g/2$

$$Z(z' + m\lambda_g/2) = Z(z') \tag{2.15}$$

2.1.2.2 Reflection Coefficient

The ratio of reflected wave voltage (or current) at any point z on the TL to incident wave voltage (or current) is defined as the reflection coefficient of voltage (or current) [2], i.e.

$$\begin{cases} \Gamma_{\mathrm{u}} = \dfrac{U_-(z)}{U_+(z)} \\ \Gamma_{\mathrm{i}} = \dfrac{I_-(z)}{I_+(z)} \end{cases} \tag{2.16}$$

Usually, the reflection coefficient of voltage (or current) is written as $\Gamma(z)$. From Eqs. (2.6) and (2.7), we can get

$$\Gamma(z) = \frac{A_2 \mathrm{e}^{-\mathrm{j}\beta z}}{A_1 \mathrm{e}^{+\mathrm{j}\beta z}} = \frac{Z_l - Z_0}{Z_l + Z_0} \mathrm{e}^{-\mathrm{j}2\beta z} = \Gamma_l \mathrm{e}^{-\mathrm{j}2\beta z} \tag{2.17}$$

Here, $\Gamma_l = \frac{Z_l - Z_0}{Z_l + Z_0} = |\Gamma_l| \mathrm{e}^{\mathrm{j}\varphi_l}$, which is called the terminal reflection coefficient. Equation (2.17) informs the relation between the terminal reflection coefficient and the input reflection coefficient at any point. The reflection coefficient has the following properties,

(1) The modulus of reflection coefficient is an invariant in lossless transmission line system

$$|\Gamma(z)| = |\Gamma_l| \tag{2.18}$$

(2) The reflection coefficient is periodical

$$\Gamma(z + m\lambda g/2) = \Gamma(z) \tag{2.19}$$

2.1.2.3 The Input Impedance and the Reflection Coefficient

From Eqs. (2.6) and (2.17), it can be concluded

$$\begin{cases} Z_{in}(z) = Z_0 \dfrac{1 + \Gamma(z)}{1 - \Gamma(z)} \\ \Gamma(z) = \dfrac{Z_{in}(z) - Z_0}{Z_{in}(z) + Z_0} \end{cases} \tag{2.20}$$

2.1.2.4 Standing-Wave Ratio

The voltage standing wave ratio (VSWR) is defined as the ratio of the voltage amplitude of the wave antinode to the voltage amplitude of the wave node on the transmission line, which is expressed by ρ

$$\rho = \frac{|U|_{\max}}{|U|_{\min}} \tag{2.21}$$

Here, $|U|_{max} = |U_+| - |U_-|$, $|U|_{\min} = |U_+| - |U_-|$, and take the Eq. (2.21) into Eq. (2.16), then the relation of ρ and Γ_l can be given

$$\begin{cases} \rho = \dfrac{1 + |U_-|/|U_+|}{1 - |U_-|/|U_+|} = \dfrac{1 + |\Gamma_l|}{1 - |\Gamma_l|} \\ |\Gamma_l| = \dfrac{\rho - 1}{\rho + 1} \end{cases} \tag{2.22}$$

2.1.3 *The Short State of the Lossless TL*

The short state, $Z_l = 0$, $\Gamma_l = -1$ means the full reflection state. For $\Gamma_l = U_l^-/U_l^+$, it means $U_l^- = -U_l^+$, so

$$\begin{cases} U(z') = U_l^+ e^{j\beta z'} + U_l^- e^{-j\beta z'} = U_l^+ e^{j\beta z'} - U_l^+ e^{-j\beta z'} = j2U_l^+ \sin \beta z' \\ I(z') = I_l^+ e^{j\beta z'} + I_l^- e^{-j\beta z'} = I_l^+ e^{j\beta z'} + I_l^+ e^{-j\beta z'} = 2I_l^+ \cos \beta z' \end{cases} \tag{2.23}$$

Then the voltage and current on the TL distributed in standing wave, and its impedance is

$$Z(z') = U(z')/I(z') = jZ_0 \tan \beta z' \tag{2.24}$$

2.2 The Basic Theory of Coaxial Wave-Guide

The coaxial waveguide is a kind of double conductor guided wave system composed of inner and outer conductors, also known as coaxial line. As shown in the Fig. 2.3. The inner and outer conductors are ideal conductors, the inner radius of the inner conductor is a, the inner radius of the outer conductor is b. The cavity between the inner and outer conductors is filled with ideal medium with permeability ε and permittivity μ. The coaxial line is a double conductor waveguide, and it can not only propagate TEM wave, but also TE wave and TM wave.

2.2.1 The Main Mode of Coaxial Transmission Line—TEM

There is the inner conductor in the coaxial transmission line (CTL), the radius of the in conductor is a, and the out radius of the medium is b. And TEM is the main mode in CTL [3].

In CTL, $k_c = 0$, so

$$\begin{cases} \nabla_t^2 \vec{E} = 0 \\ \nabla_t^2 \vec{H} = 0 \end{cases} \tag{2.25}$$

Here, the work region in the CTL is

$$\begin{cases} a \le r \le b \\ 0 \le \varphi \le 2\pi \end{cases} \tag{2.26}$$

In the cylindrical coordinate system, the Eq. (2.25) can be written as

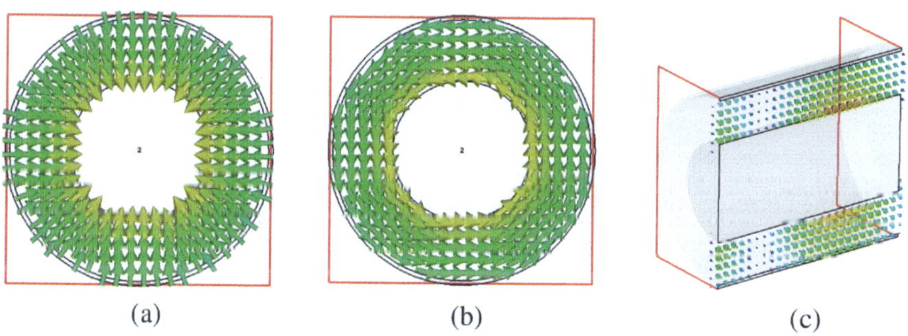

Fig. 2.3 TEM mode electromagnetic field distribution in the coaxial waveguide. **a** E-field in the cut plane, **b** H-field in cut plane, and **c** H-field in cut plane

$$\begin{cases} \frac{1}{r}\frac{\partial}{\partial r}(r\frac{\partial \vec{E}}{\partial r}) + \frac{1}{r^2}\frac{\partial^2 \vec{E}}{\partial \varphi^2} = 0 \\ \frac{1}{r}\frac{\partial}{\partial r}(r\frac{\partial \vec{H}}{\partial r}) + \frac{1}{r^2}\frac{\partial^2 \vec{H}}{\partial \varphi^2} = 0 \end{cases} \tag{2.27}$$

For the symmetrical structure in CTL, $\partial \vec{H}/\partial \varphi = 0$ and $\partial \vec{E}/\partial \varphi = 0$, so

$$\begin{cases} \frac{1}{r}\frac{\partial}{\partial r}(r\frac{\partial \vec{E}}{\partial r}) = 0 \\ \frac{1}{r}\frac{\partial}{\partial r}(r\frac{\partial \vec{H}}{\partial r}) = 0 \end{cases} \tag{2.28}$$

And the general solution can be written by the potential function

$$\vec{E} = \nabla\Phi(r,\varphi)e^{-j\beta z} = \vec{E}_t(r,\varphi)e^{-j\beta z} \tag{2.29}$$

So,

$$-\nabla\left[\frac{1}{r}\frac{\partial}{\partial r}(r\frac{\partial \Phi}{\partial r}) + \frac{1}{r^2}\frac{\partial^2 \Phi}{\partial \varphi^2}\right] = 0 \tag{2.30}$$

Considering the symmetry, $\frac{\partial}{\partial \varphi} \to 0$, and the arbitrariness to select potential function, it can be given

$$\begin{cases} \frac{1\partial}{r\partial r}\left(\frac{r\partial \Phi}{\partial r}\right) = 0 \\ \Phi = A\ln r + B \end{cases} \tag{2.31}$$

So, the field can be written

$$\begin{cases} E_r = E_0\frac{a}{r}e^{-j\beta z} \\ H_\varphi = \frac{E_0 a}{\eta r}e^{-j\beta z} \end{cases} \tag{2.32}$$

And the corresponding field distribution is shown in Fig. 2.3.

2.2.2 *The Parameters of the Main Mode in CTL*

The coaxial current and the voltage between the inner and outer conductors are defined as

$$\begin{cases} I = \oint_l H_\varphi dl = \int_0^{2\pi} H_\varphi r d\varphi = 2\pi r H_\varphi |_{r=a} = \dfrac{2\pi E_0 a}{\eta} e^{-j\beta z} \\ U = \int_a^b E_r dr = E_0 a \ln(\dfrac{b}{a}) e^{-j\beta z} \end{cases} \tag{2.33}$$

Characteristic impedance

$$Z_0 = \frac{U}{I} = \frac{60}{\sqrt{\varepsilon_r}} \ln(\frac{b}{a}) \Omega \tag{2.34}$$

Attenuation

$$\alpha_c = \frac{8.686 R_s \sqrt{\varepsilon_r}(1 + \frac{b}{a})}{2\pi b(120 \ln \frac{b}{a})} \text{dB/m} \tag{2.35}$$

Power capacity

$$P_{\max} = \frac{\sqrt{\varepsilon_r}}{120} E_{\max}^2 a^2 \ln(\frac{b}{a}) W \tag{2.36}$$

The parameters of TEM wave in CTL are in following.
Cutoff Wavenumber

$$k_c = \sqrt{\gamma^2 + k^2 = 0} \tag{2.37}$$

Cutoff Wavelength

$$\lambda_c = \infty \tag{2.38}$$

Phase constant

$$\beta = \omega \sqrt{\mu \varepsilon} \tag{2.39}$$

Phase velocity

$$v_p = \frac{\omega}{\beta} = \frac{1}{\sqrt{\varepsilon \mu}} \tag{2.40}$$

Wave impedance

$$Z_{TEM} = \frac{E_r}{H_\varphi} = \sqrt{\frac{\mu}{\varepsilon}} = \eta \tag{2.41}$$

2.2.3 High Order Modes in the Coaxial Transmission Line (CTL)

In the real, CTL works in TEM mode. But there will be some high order modes: TM and TE when the frequency is very high.

The TM mode and TE mode in CTL can be analyzed similarly as in cylindrical wave-guide. Under the given boundary condition solve the wave equation of E_z and H_z, so that get the field distribution of TE_{mn} and TM_{mn}, and cutoff wavelength λ_{cmn}.

The character equation of TE mode in CTL is

$$\lambda_{cTE_{m1}} \approx \frac{\pi(b+a)}{m}(m = 1, 2, \nu) \tag{2.42}$$

$$\begin{cases} \lambda_{cTE_{11}} \approx \pi(a+b) \\ \lambda_{cTE_{01}} \approx 2(b-a) \end{cases} \tag{2.43}$$

The character equation of TM mode in CTL is

$$\frac{J_m(k_c a)}{J_m(k_c b)} = \frac{N_m(k_c a)}{N_m(k_c b)} \tag{2.44}$$

$$\lambda_{cTM_{01}} \approx 2(b-a) \tag{2.45}$$

The cutoff wavelength of these modes in CTL are shown in Fig. 2.4.

2.3 The Basic Theory to Slot Antenna

XOZ is the infinite PEC plane, and there is a slot along z axis with length $2l$, width w ($w << \lambda$) on the plane. For any excitation mode to the slot, the E-field in the slot always is vertical to the length direction of the slot, and the E-field is symmetrically distributed up and down to the midpoint of the joint, as shown in Fig. 2.5a [4].

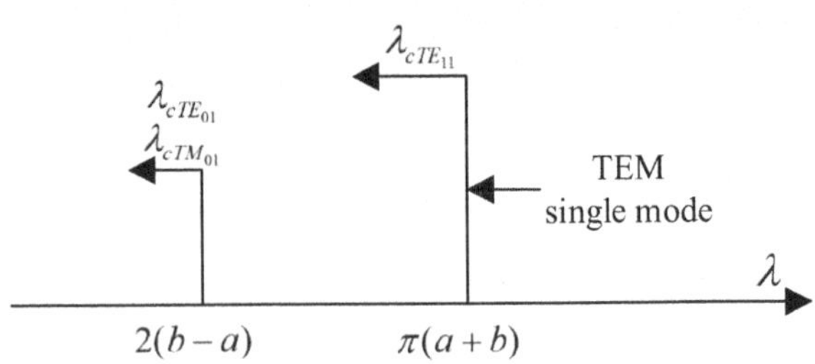

Fig. 2.4 Cutoff wavelength of each mode in the coaxial waveguide

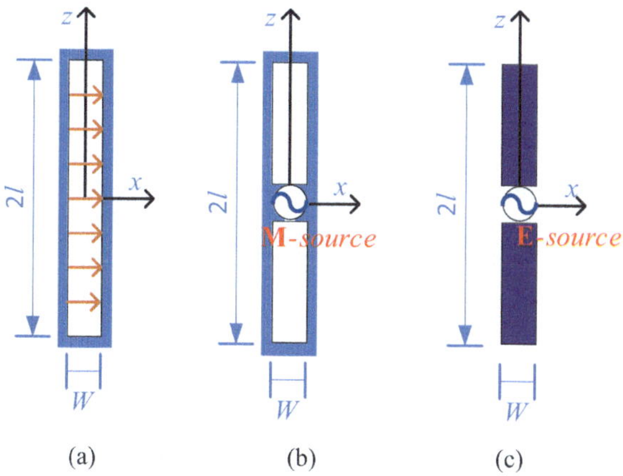

Fig. 2.5 Ideal slot antenna and tabular dipole, **a** E-field distribution in the slot, **b** Magnetic source to excite the slot and **c** the Electronic source to excite the dipole

For $\vec{J}_m = -\hat{n} \times \vec{E}$, the equivalent magnetic current is reversal on two surfaces inside and outside the slot [5, 6].

If the field of the ideal slot antenna is the same as the one excited by the magnetic current source in the $y > 0$ half space, it will be minus sign in the $y < 0$ half space. Because on the same surface, the equivalent magnetic current is symmetrically distributed up and down to the midpoint of the slot, the ideal slot antenna can be equivalent to a symmetrical slot excited by the magnetic current source, as shown in Fig. 2.5b. According to Babinet's principle, it is obvious that the complementary one is a plate-shaped symmetric oscillator of the same size, as shown in Fig. 2.5c.

The far field of the planar dipole is same as the thin linear dipole, that is

$$E_\theta = \mathrm{j}60 I_m \frac{\cos(kl\cos\theta) - \cos kl}{\sin\theta} \frac{\mathrm{e}^{-\mathrm{j}kr}}{r} \tag{2.46}$$

In the equation, θ is the angle between the line direction and the dipole. It means the pattern function of the dipole is same as the idea slot antenna.

$$f_\theta = \frac{\cos(kl\cos\theta) - \cos kl}{\sin\theta} \tag{2.47}$$

For idea half wavelength slot, $2l = \lambda/2$, and

$$f_\theta = F_\theta = \frac{\cos(\frac{\pi}{2}\cos\theta)}{\sin\theta} \tag{2.48}$$

The pattern of the idea slot antenna is same as the diploe with same length, but their E-plane and H-plane are exchanged according to the duality principle.

2.4 Basic Theory of Array Antenna

2.4.1 Product Theorem of Pattern

Array antenna is a kind of special antenna form which is composed of several antenna units, which are arranged according to certain geometric rules. Compared with a single antenna, the array antenna can obtain the desired radiation characteristics through appropriate cell selection, arrangement and excitation settings, including the direction and maximum gain of the pattern beam. For the N separate same antenna unit arrays shown in Fig. 2.6, the direction function $\vec{f}(\theta, \varphi)$ in the direction (θ, φ) can be expressed as

$$\vec{f}(\theta, \varphi) = \sum_{n=1}^{N} I_n e^{jk\vec{r}_n\vec{r}} \vec{f}_n(\theta, \varphi) \tag{2.49}$$

Here, $I_n = |I_n|e^{ja_n}(n = 1, 2, \nu N)$ is the complex excitation coefficient of the nth element, a_n is the excitation phase of the nth element, $k = 2\pi/\lambda$ is the wave number, $\vec{r}_n = x_n\hat{x} + y_n\hat{y} + z_n\hat{z}(n = 1, 2, \nu N)$ is the unit vector at field point, and $\vec{f}_n(\theta, \varphi)$ $(n = 1, 2, \nu N)$ is the direction function of the nth element.

If all the elements are same, the direction function can be expressed by $\vec{f}_0(\theta, \varphi)$, so Eq. (2.44) can be rewritten as

$$\vec{f}(\theta, \varphi) = \vec{f}_0(\theta, \varphi) \sum_{n=1}^{N} I_n e^{jk\vec{r}_n\vec{r}} = \vec{f}_0(\theta, \varphi)\vec{f}_a(\theta, \varphi) \tag{2.50}$$

Here, $|\vec{f}_0(\theta, \varphi)|$ is the element pattern function, which is only decided by geometry structure, size of the antenna, and is called the element factor. $|\vec{f}_a(\theta, \varphi)|$ is related to

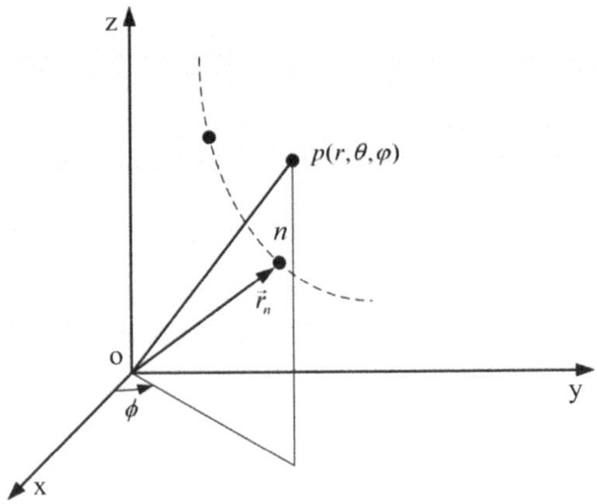

Fig. 2.6 N element antenna array with arbitrary arrangement

the current distribution on element, array space distribution and number of elements, which is called array factor.

In Eq. (2.50), if all the elements are same, the array pattern is just the product of the element factor and array factor, this is just called the product theorem of pattern [7–9].

2.4.2 Uniform Linear Array

The uniform linear array (ULA) is a linear array with equal spacing and equal amplitude (equal amplitude distribution) of the current of each element, and the phase increases in equal order (linear phase distribution). As shown in Fig. 2.7, N elements are arranged along the z-axis, the interval between the two adjacent elements is d, and the phase difference of the excitation current is β, then the radiation field is

$$E = E_0 \sum_{i=1}^{n} e^{j(i-1)\psi} \tag{2.51}$$

Here, E_0 is the radiation field of the 1st idea point source.

$$\psi = kd\cos\theta + \beta \tag{2.52}$$

Take Eq. (2.47) into (2.46), and the direction function of the ULA can be described as:

$$|f_a(\theta)| = \left|\sum_{i=1}^{n} e^{j(i-1)\psi}\right| = \left|\frac{\sin(\frac{n}{2}\psi)}{\sin(\frac{\psi}{2})} e^{j(n-1)\psi/2}\right| = \left|\frac{\sin(\frac{n}{2}\psi)}{\sin(\frac{\psi}{2})}\right| \tag{2.53}$$

The maximum of Eq. (2.48) appeared at $\psi = 2m\pi(m = 0, \pm 1, \pm 2\nu, \ldots)$, it can be obtained by using the law of Lobida

$$f_{aM} = \left|\frac{\sin(\frac{n}{2}\psi)}{\sin(\frac{\psi}{2})}\right|_{\psi=2m\pi} = n \tag{2.54}$$

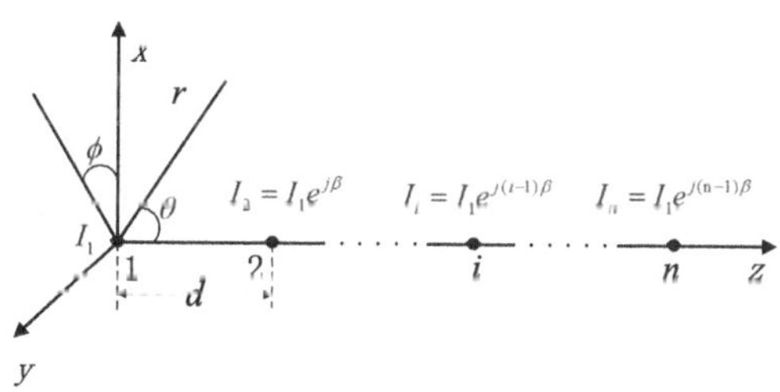

Fig. 2.7 N element uniform linear arrays

The normalized direction function of the ULA is

$$|F_a(\theta)| = \frac{1}{n}\left|\frac{\sin(\frac{n}{2}\psi)}{\sin(\frac{\psi}{2})}\right| \tag{2.55}$$

ψ is the phase difference of the radiation field of two adjacent antenna elements. Equation (2.47) shows that ψ depends on two factors, one is the space phase difference of adjacent elements $kd\cos\theta$, the other is the current phase difference of adjacent elements β. It can be seen from Eq. (2.48) that the maximum value of the pattern $f_{aM} = n$ is the result of equal amplitude current of each element when $\psi = 2m\pi$. It usually required that there is only one maximum in the main beam at $\psi = 0$ to the array. Let the direction of the maximum value of the main lobe be θ_M, it can be obtained from Eq. (2.47)

$$\beta = -kd\cos\theta_M \tag{2.56}$$

When the current of each element lags the phase $kd\cos\theta_M$ in turn, the space phase difference at the direction of θ_M is just compensated by the current lag phase difference, so this direction becomes the maximum radiation direction of the antenna array. The direction of the beam can be controlled by controlling the phase of the excitation source. If it needs the end fire beam, $\theta_M = 0°$ or $180°$, or the side beam at $\theta_M = 90°$.

Side beam array is a linear array whose maximum radiation direction is perpendicular to the array line $\theta_M = 90°$. According to Eq. (2.51), the current phase difference $\beta = 0$ of adjacent elements of the side beam array is a linear array with same phase. In the maximum radiation direction of the side radiation array, there is no wave path difference between the distance from each element to the observation point, so the current of each element needn't phase difference. For a side radiation array,

$$\psi = kd\cos\theta \tag{2.57}$$

$$F_a(\theta) = \frac{\sin(\frac{n}{2}kd\cos\theta)}{n\sin(\frac{1}{2}kd\cos\theta)} \tag{2.58}$$

2.4.3 Grating Lobe and the Interval

To Eq. (2.47), in the available range of θ ($0° \le \theta \le 180°$), the corresponding range of ψ is $\langle -kd + \beta, kd + \beta \rangle$, and we called this range of θ and ψ as visible range of the antenna array. In mathematics, if ψ is out of the above range, the corresponding θ is not in the visible region of space. The ψ range corresponding to the visible region of θ is called the visible region of ψ.

In fact, the visible region of ψ is decided by the elements spacing d. If d is too large, there will be multi beams with same maximum values in the pattern. And these maximum values appear at $\psi = 2m\pi$. To the lobe with the maximum at $\psi = 0$ is called the main lobe of the pattern, and other lobes with maximum values are called grating lobes. Generally, the grating lobes is harmful, and the way to eliminate the gating lobe is to choose the right interval, so that it does not exceed a certain limit value. Only if ψ is in the visible range and isn't equal to 2π, there isn't grating lobe in the array pattern.

$$\begin{cases} -kd + \beta > -2\pi \\ kd + \beta < 2\pi \end{cases} \tag{2.59}$$

To solve this joint inequality group, and get

$$\frac{d}{\lambda} < 1 - \frac{|\beta|}{2\pi} \tag{2.60}$$

This equation is just the interval condition to eliminate the grating lobe. Even though the interval d satisfies Eq. (2.54), the visible region of ψ may be close to the grating lobe region, so that the side lobe of the pattern become large.

To eliminate the whole grating lobe besides the maximum of the grating lobe, it should have

$$\begin{cases} -kd + \beta \geq -(2\pi - \Delta\psi_0) \\ kd + \beta \leq 2\pi - \Delta\psi_0 \end{cases} \tag{2.61}$$

Here, $\Delta\psi_0 = 2\pi/n$, is the half of the zero power width of the general pattern grating lobe, so

$$\frac{d}{\lambda} \leq 1 - \frac{1}{n} - \frac{|\beta|}{2\pi} \tag{2.62}$$

Take the β of the uniform side radiation array into Eq. (2.62), and get the interval condition to cancel the whole grating lobe as:

$$\frac{d}{\lambda} < \frac{n-1}{n} \tag{2.63}$$

2.5 Summary

The purpose of this book is to study the slot array high gain omnidirectional circular polarization antenna (OCPA) based on coaxial waveguide transmission line. In order to better guide the design of such a high gain OCPA through theoretical knowledge, this chapter mainly introduces the basic theoretical knowledge involved in the antenna design, and then introduces the theories of transmission line, coaxial waveguide, slot antenna and array antenna, although there are not only these basic theories involved in the design of OCPA in this book. Through the introduction of these basic theories, it will support theoretical basis to the later analysis of the antenna design principle make the description of the antenna more clear, and better to understand the OCPA and its extended application design.

References

1. C. Liang, Y. Xie, B. Guan, et al., *Concise Microwave* (Higher Education Press, 2006)
2. D.M. Pozar, *Microwave Engineering*, 3rd edn. (Wiley, USA, 2004)
3. C. Xie, K. Rao, *Electromagnetic Field and Electromagnetic Wave*, 4th edn. (Higher Education Press, Beijing, 2006)
4. J.D. Kraus, R.J. Marhefka, et al., *Antenna*, 2nd edn. (Tata McGraw-Hill, 2001)
5. S. Zhong, *Antenna Theory and Technology*, 2nd edn. (Electronic Industry Press, Beijing, 2011)
6. W. Wei, *Antenna Principle* (National Defense Industry Press, Beijing, 1985)
7. P. Focardi, P.R. Brown, Y. Rahmat-Samii, *Space Antenna Handbook* (2012), pp. 250–313
8. D. Pozar, D. Schaubert, *Microwave Antennas: The Analysis and Design of Microstrip Antennas and Arrays* (IEEE Xplore, 1995)
9. R. Gabriel, M. Gottl, G. Klinger, Antenna array. US Patent 6025812 (2000)

Chapter 3
Principles of Omnidirectional CP Slot Array Antenna

3.1 Introduction

Omnidirectional circularly polarized (CP) antenna (OCPA) is the key to realize 360° full coverage with being insensitive to the orientations of the incident waves in modern wireless communication, and it can also resist multipath interference and channel fading in the complex environment [1]. Especially, it is very useful in the radio channel in civil or military communication scenario.

Recently, there are many researchers to present and study the OCPA, and many kinds of OCPA are reported to develop the wireless communication technology. In [2–4], some planar dual-band OCPA structures were proposed, their sizes are small, and bandwidth is narrow. In [5, 6], the OCPA dielectric resonator antennas (DRAs) are introduced, which can be manufactured very small in size for their high dielectric constant, and their available impedance matching bandwidth with return loss being smaller than –10 dB is about 20% while their available CP bandwidth is only about 3–8% with the 3 dB AR. Monopole and loop are combined to construct the simple EM antenna to achieve omnidirectional CP property in [7], its available bandwidth with return loss $|S_{11}| < -10$ dB is 15%. A compact OCPA was designed in [8] with four curved monopoles, and they were excited at same time by a feeding network. In [9] a broadband omnidirectional CP antenna was proposed whose impedance bandwidth was 45%. These OCPAs' bandwidths are wide, but they can't work with a stable high gain in the whole working bandwidth.

In this chapter, a novel OCPA by cutting slots array on the outside shell of the coaxial cylinder is proposed [10]. The omnidirectional CP wave is radiated by the four orthogonal slot pairs with an interval of half wavelengths surrounding the axis of coaxial cylinder in the out conductor. Four circles of the slot pairs along the axis construct radiation slots array to realize the stable high gain property. In further, it is convenient to increase or decrease the number of the basic slot array elements along the axis to adjust the antenna beam width and gain for different application.

The antenna's geometry structure, circuit and EM principle and design method are described in Sect. 3.2. The simulated and measured results and analysis, including

J. Geng, *Omnidirectional Slots Antenna*,
https://doi.org/10.1007/978-981-15-9089-4_3

bandwidths, axial ratio and gain pattern are shown in Sects. 3.3 and 3.4. The summary are in Sect. 3.5.

3.2 Geometry Structure and Principle

3.2.1 Antenna Structure

The geometrical structure of the proposed OCPA is shown in Fig. 3.1. It is designed based on the slot array in the coaxial cylinder, and the interval cavity between the inner and outer conductor is filled by Teflon. The thickness of the cylinder outer conductor is $t = 1.5$ mm. The antenna consists of three parts, including radiation part, matching part and feeding part.

The radiation unite is orthogonal slot pair, and these orthogonal slot pairs distribute in the conductor shell of the coaxial, they are symmetrical around the coaxial wire, as presented in Fig. 3.1a, c. All of these symmetrically arranged slots are combined into slot array in Fig. 3.1.

Band. Although the characteristic impedance of the SMA is 50 Ω, but its size smaller than the coaxial cylinder. Here a tapered coaxial cylinder *L4* is introduced to connect the radiation part of the coaxial cylinder and SMA adaptor in Fig. 3.1b. Its ratio of out and inner radius is constant to realize a given characteristic impedance.

The complicated structure still limited impedance band. So, three rings *L1*, *L2* and *L3* close to the feeding port are introduced to improve the impedance matching in wide band, and they are called matching part. The size of these three coaxial cylinders *L1*, *L2* and *L3* are different with different characteristic impedance and electrical length, just like the branches parallel connecting to the transmission line.

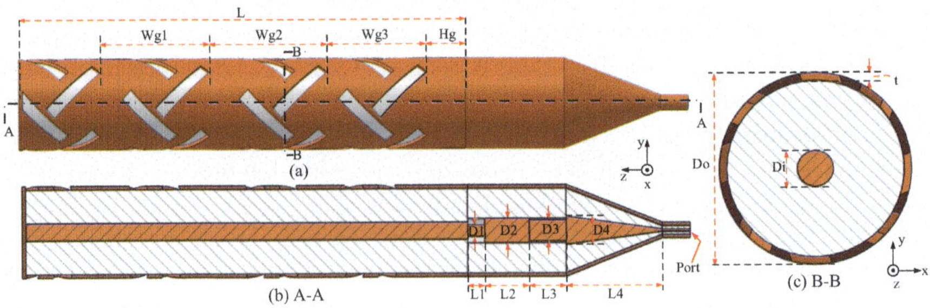

Fig. 3.1 Geometry of the omnidirectional circularly polarized (CP) antenna. **a** Side view, **b** A–A section view, **c** B–B section view [10, 11] Figure reproduced with permission from: [11], © 2015 IEEE

3.2.2 Design Principle to Omnidirectional CP Radiation Performance

The geometrical structure of the slot pairs is shown in Fig. 3.2. The slot distributes on the out conductor shell with angle of 45° or −45° from the axis of the coaxial cylinder, and all the slots are orthogonal each other. Two adjacent slots combine a pair in Fig. 3.2, and the center distance between these two slots is $\lambda_g/4$ (λ_g is the medium wavelength) along the axis. So, the radiated EM field from the two slot is orthogonal with each other and their phase difference is 90° for feeding along z axis. That is just the reason to CP radiation, and the RHCP property is realized in [12].

In theory, the slot can be regarded as magnetic current, of which the length of the slot is $\lambda_g/2$ in the fed direction. As shown in Fig. 3.2, the slot is rotated an angle of 45° from the fed direction, and the length of the slot is designed $L_s = \lambda_g/2$.

In Fig. 3.1b, there are four slot pairs uniformly distributed in the outer conductor shell around the coaxial cylinder, it is just a basic omnidirectional circular polarization unit. Four rounds of the basic omnidirectional circular polarization units are arranged uniformly along the axis of the coaxial cylinder to get high gain, as shown in Fig. 3.1c.

In Fig. 3.1, only the EM field radiated from these rounds slots with same phase, the main beam can be omnidirectional in the azimuth plane with high gain. So, the interval of four round units is set $W_g = \lambda_g$ with each other. Here, the medium filled in the cylinder is Teflon with a permittivity of $\varepsilon_r = 2.1$, and the calculated medium wavelength is $\lambda_g = 0.7\lambda_0$ (λ_0 is the space wavelength), which is the interval between the slot array units too. $0.7\lambda_0$ is just corresponding to high radiation efficiency of the array aperture. In another side, as shown in Fig. 3.2b, the last element of the array is cut an angle of D_g from other element, which can adjust the impedance matching of the shortened terminal, and decrease the reflect wave in the end. So, the

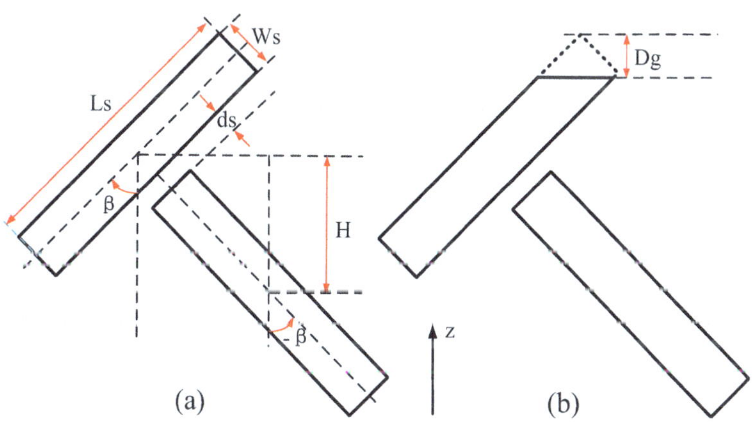

Fig. 3.2 Geometry of slot pairs on out conductor of the coaxial cylinder. **a** Normal slot pairs. **b** Changed slot pairs. Figure reproduced with permission from: [11], © 2015 IEEE

EM wave runs in the coaxial cylinder with traveling wave and radiates from these slots simultaneously with high efficiency.

All above improvements enable the slots array to perform omnidirectional radiation in wide CP band and impedance bandwidth with high gain.

Here, the antenna is fed in the axial direction. The interval along axis between these two orthogonal slots is $H = \lambda_g/4$, so that the phase difference between two slots is 90°. As we know, the CP wave can be decomposed into two linearly polarized waves whose polarization directions are perpendicular to each other and the phase difference is 90°. So, the adjacent two orthogonal slots on the outside can radiate the CP wave.

Since the polarization direction of the electric field radiated by each slot is perpendicular to the slot, we take out a pair of mutually perpendicular slots, and establish a rectangular coordinate system on the slot. And the x-axis and y-axis of the established coordinate system are consistent with the polarization direction of the electric field of these two gaps, and the wave propagation direction is along the + z axis [13, 14], as shown in Fig. 3.3.

Assume the total electric field of these two slots is $\vec{E}$, the electric field of the 1st slot is $\vec{E}_1$, $\vec{E}_2$ to the 2nd slot. So, the relation between these electric fields are given as

$$\vec{E} = \vec{E}_1 + \vec{E}_2 = E_x\vec{a}_x + E_y\vec{a}_y \tag{3.1}$$

Here, E_{x0} ($E_{x0} > 0$) is the magnitude of E_x, E_{y0} ($E_{y0} > 0$) is the magnitude of E_y, and the phase of E_y lags behind E_x by φ (if the phase of E_y is ahead of E_x, then $\varphi < 0$), so Eq. (3.1) can be written as

$$\vec{E} = E_{x0}e^{-jkz}\vec{a}_x + E_{y0}e^{-jkz}e^{-j\varphi}\vec{a}_y \tag{3.2}$$

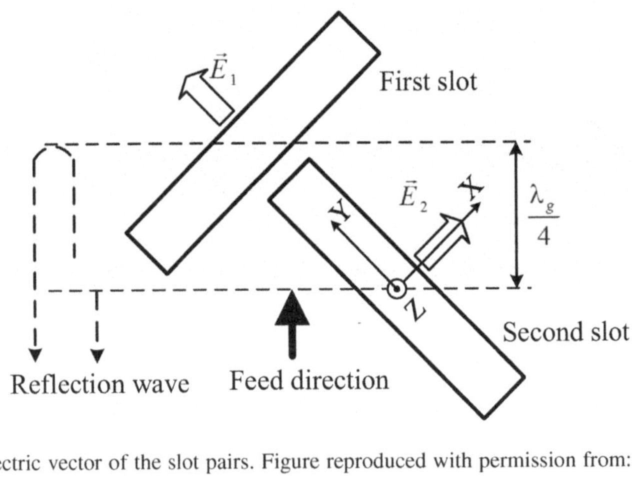

Fig. 3.3 Electric vector of the slot pairs. Figure reproduced with permission from: [11], © 2015 IEEE

Its instantaneous expression is

$$\vec{E}(z,t) = E_{x0}\cos(\omega t - kz)\vec{a}_x + E_{y0}\cos(\omega t - kz - \varphi)\vec{a}_y \tag{3.3}$$

Here,

$$\begin{cases} E_x(z,t) = E_{x0}\cos(\omega t - kz) \\ E_y(z,t) = E_{y0}\cos(\omega t - kz - \varphi) \end{cases} \tag{3.4}$$

Cancel $\cos(\omega t - kz)$ from Eq. (3.4), and can get the relation of $E_x(z,t)$ and $E_y(z,t)$:

$$\left[\frac{E_x(z,t)}{E_{x0}}\right]^2 + \left[\frac{E_y(z,t)}{E_{y0}}\right]^2 - \frac{2E_x(z,t)E_y(z,t)}{E_{x0}E_{y0}}\cos\varphi = \sin^2\varphi \tag{3.5}$$

Here, the spacing between these two slots is $\lambda_g/4$, so their phase difference is $\varphi = 90°$. If we set the magnitude of these two slots as $E_{x0} = E_{y0} = E_0$, the Eq. (3.5) can be written as

$$E_x^2(z,t) + E_y^2(z,t) = E_0^2 \tag{3.6}$$

In Eq. (3.4), the magnitude of $\vec{E}(z,t)$ doesn't vary with time t,so the angle between the direction of $\vec{E}(z,t)$ and +x can be written

$$\begin{aligned} \alpha &= \arctan\frac{E_0\cos(\omega t - kz \pm \frac{\pi}{2})}{E_0\cos(\omega t - kz)} \\ &= \arctan[\mp tan(\omega t - kz)] = \mp(\omega t - kz) \end{aligned} \tag{3.7}$$

In Eq. (3.7), for any z, the direction of $\vec{E}(z,t)$ rotates with the constant angular frequency ω with time, this is the circular polarization wave. The feeding direction is shown in Fig. 3.3, the phase of E_y lags behind E_x by $\varphi = 90°$, $\alpha = \omega t$, so the electric field vector rotating angle increases along the anticlockwise direction with time, this is the right hand circular polarization (RHCP) wave. Conversely, if the feeding direction is reversed, the phase of E_x lags behind E_y by $\varphi = 90°$, the electric field vector rotating angle increases along the clockwise direction with time, this is the left hand circular polarization (LHCP) wave.

Here the feeding direction of this antenna in Fig. 3.3 results in $\varphi = 90°$, so it is RHCP omnidirectional antenna. Because our antenna is axial symmetrical antenna, the feeding port and short terminal can be easily exchanged to become LHCP omnidirectional antenna.

In fact, the electric fields of these two slots are not always equal, $E_{x0} \neq E_{y0}$, so that the real radiation wave is not the standard circular polarization wave, but an elliptical polarization wave.

For the slot pair as shown in Fig. 3.3, as the spacing between the two gaps is $\lambda_g/4$ along the axis, the position difference of the reflected wave when passing through the first slot and the second slot is $\lambda_g/2$, and the corresponding phase difference is 180°. Taking the first slot as the reference plane, the reflected wave overlaps and cancels each other in the reverse direction on the reference plane. Therefore, such slot structure is easy to better reflection coefficient of the antenna.

Four pairs of slots are symmetrically distributed around the coaxial axis on the outer wall of the coaxial to achieve omnidirectional radiation. In order to increase the gain of the antenna, a circle of slots is taken as an omni-directional circular polarization unit of the antenna, and four such units are designed on the outer wall along the coaxial line to form an array as shown in Fig. 3.1a. The antenna is fed along the axis direction, which means that the feed mode of the array is equivalent to the serial feed. According to the formula (2.51) of the basic theory of the array antenna in Chap. 2, in order to make the synthesized pattern of the array to be perpendicular to the array direction, the feed phase of each unit should be same. The four omnidirectional circularly polarized units of the antenna are distributed along the axial direction, and the spacing between the units is λ_g to ensure that the four units are fed in same phase, so that the pattern direction is perpendicular to the array axis.

The dielectric between the inner and outer conductors of the coaxial line is Teflon with dielectric constant $\varepsilon_r = 2.1$. One reason to fill media in coaxial lines is to reduce the antenna size. Secondly, only if the feeding phase of these four circle slots units are same or with difference 360°, the slots array will achieve high gain. To suppress the grating lobe, the interval between these units must satisfy Eq. (2.60). Due to the permittivity, according to formula (2.56), the spacing between array elements meets the requirements of formula (2.60) for gate lobe suppression.

Another important reason is that in order to design the antenna with high gain, it must be realized by array. According to the basic theory of array in Chap. 2, in order to suppress the grid lobes brought by array, the array spacing and free space wavelength must meet the requirements of Eq. (2.60). Due to the permittivity $\varepsilon_r = 2.1$, according to formula (2.60), the spacing $\lambda_g = \lambda_0/\sqrt{\varepsilon_r} = 0.69\lambda_0$ between array elements can meet the requirements of formula (2.60) for grating lobe suppression.

The 3D simulation patterns at 5.5 GHz are shown in Fig. 3.4 when the OCPA is filled with medium or not. Figure 3.4a is the pattern when the coaxial line is filled with Teflon medium. The design spacing of the four elements is $0.69\lambda_0$, and the main lobe of the pattern points to the horizontal direction, which is the desired design. Figure 3.4b corresponds to the pattern when the coaxial line is not filled with medium. The design spacing of the four elements is λ_0. It is obvious that the pattern of the antenna has large distortion and grating lobes, which should be avoided.

As shown in Fig. 3.2b, the last circle slots near the closed short-circuit end is not the same as the other three circles lots. Comparing with the normal slots, this circle slots is cut to a part of D_g. As the overall structure of the antenna is a coaxial transmission line with short circuit at the end, according to the short circuit state of lossless transmission line in Chap. 2, the short circuit at the end will cause the total reflection of electromagnetic field to form standing wave. Here, the antenna we designed is

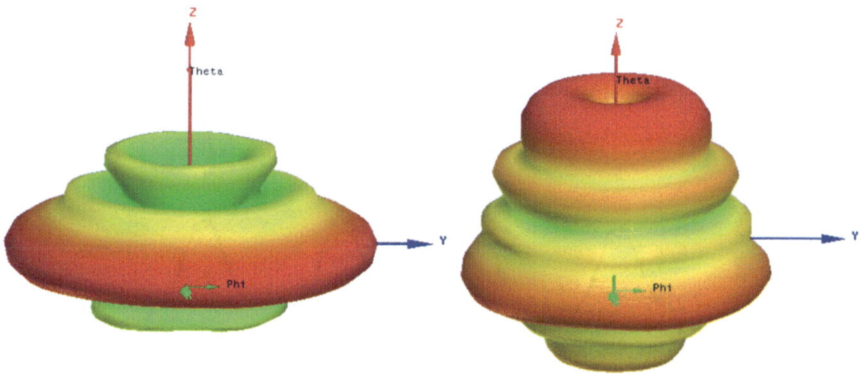

(a) (b)

Fig. 3.4 Radiation patterns of the omnidirectional CP antenna at 5.5 GHz when filling Teflon (**a**) and filling air (**b**) between inner and outer conductor [10]

based on the traveling wave structure. According to the analysis of the circular polarization principle of the antenna in the previous part, if the electromagnetic wave is reflected from the end to the feed end, it will radiate the left-handed circular polarization wave, which will seriously affect the circular polarization performance of the antenna, and the axial ratio parameter will be very poor. Therefore, the end circle slot is designed to be different from the normal slot. In fact, there are two purposes: one is to improve the gain of the omnidirectional circularly polarized antenna as an array element like other slots; In another side, this circle slots is to change the resistance of the last circle slots by optimizing the D_g parameter, so that the coaxial transmission line with short circuit at the end is equivalent to one under the action of cutting the irregular shape gap. The end matched coaxial transmission line can reduce the reflection of electromagnetic wave caused by the end mismatch. In fact, the special slots in the last circle not only plays the role of array radiation, but also adjust the antenna impedance matching. In the later period, when the antenna parameters are optimized and adjusted, the D_g parameter can optimize the circular polarization performance and the axial ratio of the antenna.

Figure 3.5 shows the axial ratio pattern comparison in the horizontal plane at 5.5 GHz between the antenna with last circle of slot angle cutting and not cutting. And the legend part, together with the text description, gives the three-dimensional simulation model of the antenna at these two cases. In Fig. 3.5, the red curve represents the axial ratio result when the last circle slot is chamfered, and the blue curve represents the axial ratio result when the last circle of the slot is complete. It can be seen that the axial ratio of the antenna in the horizontal plane at the frequency point of 5.5 GHz is less than 3 dB, and the circular polarization performance of the antenna is good. However, when the angle is not cut, the axial ratio of the antenna at this frequency point deteriorates, and the axial ratio in the whole horizontal plane ranges from 3 dB to 6 dB, and it means the circular polarization performance of the antenna is deteriorated.

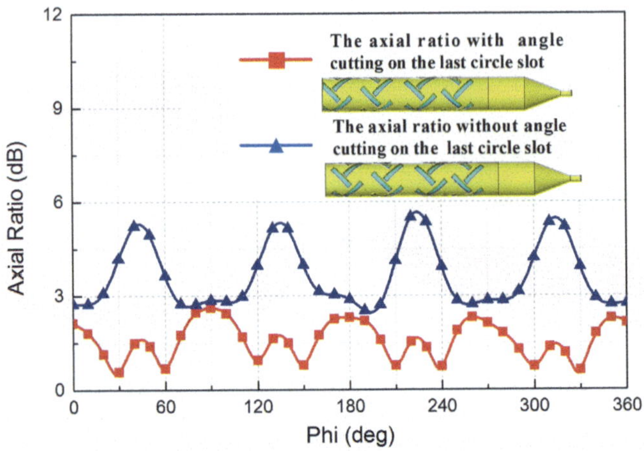

Fig. 3.5 Axial ratio pattern on omnidirectional plane (x–y plane) of the omnidirectional CP antenna at 5.5 GHz when the last round of the slot pairs being cut a part and being normal

3.2.3 *Impedance Matching and Feed Design*

After the preliminary design, the antenna structure is as shown in Fig. 3.6. In the simulation, the antenna is fed directly by setting the wave port. Considering the actual feeding of the antenna, the standard SMA connector is used for antenna feeding. The SMA connector is coaxial structure, and the antenna designed in this section is also based on coaxial structure, so the coaxial gradient line is used here to directly connect the large-scale coaxial line of the antenna to the standard SMA connector through transformation, as shown in Fig. 3.7. The return loss of the antenna S11 obtained by simulation is shown in Fig. 3.8, which is far from the general requirements. In order to improve the return loss of the antenna and meet the general requirements of the antenna design, it is necessary to further design the antenna matching structure to make in band.

Fig. 3.6 Feeding on the simulation model of the omnidirectional CP antenna

Fig. 3.7 Actual feeding model of the omnidirectional CP antenna

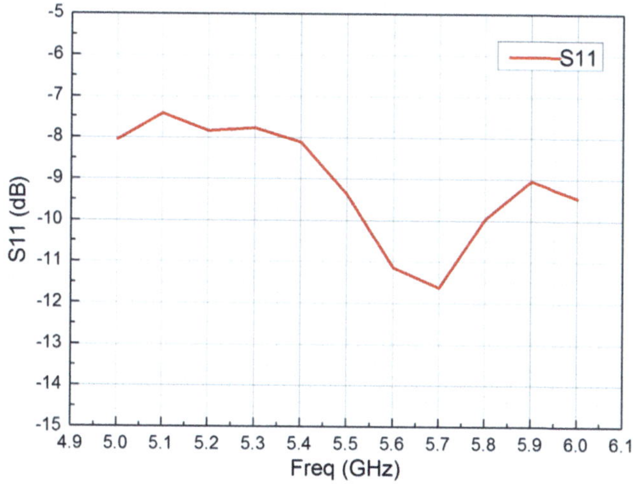

Fig. 3.8 The S-parameter results of the omnidirectional CP antenna

The matching structure is designed and inserted between the radiation part and the feed part. In the feed port plane of the radiation part of the antenna as shown in Fig. 3.6, the input impedance of the antenna varying with the frequency is shown in Fig. 3.9. It can be seen that the impedance of the antenna changes with the frequency, and the characteristic impedance of the SMA connector used is 50 Ω. It is impossible to match the impedance of each frequency point in the required frequency band to 50 Ω. In order to meet the requirements of the antenna for the return loss parameters in

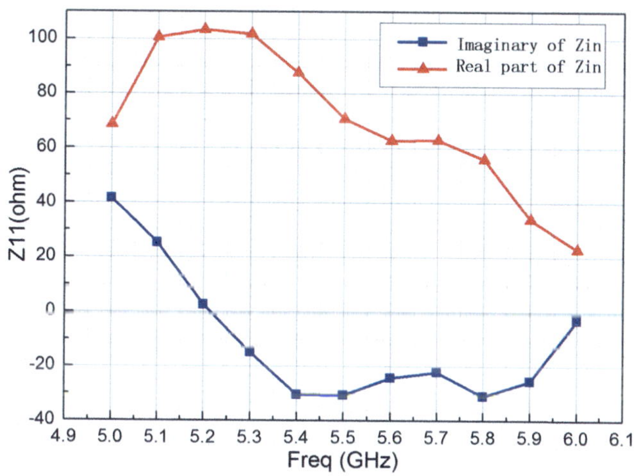

Fig. 3.9 Impedance results of the omnidirectional CP antenna. Figure reproduced with permission from: [11], © 2015 IEEE

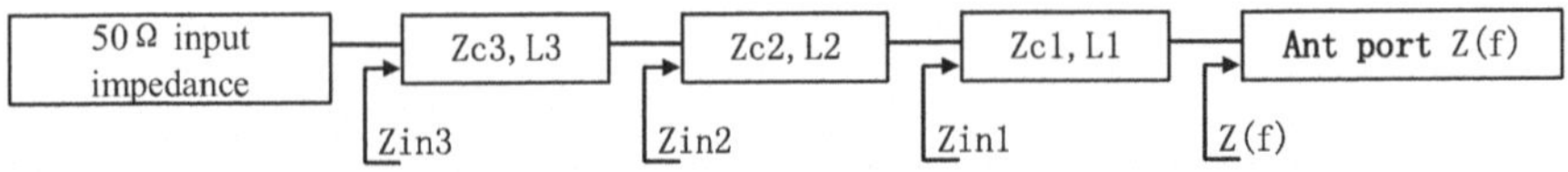

Fig. 3.10 The impedance match principle model of the omnidirectional CP antenna

a given frequency band, the matching part is designed based on matching the lowest, middle and highest frequency points in a given frequency band to 50 Ω.

The antenna is coaxial structure, and the feed structure includes three pieces of coaxial line with different characteristic impedance and electrical length. The impedance matching model is shown in Fig. 3.10. Here, $Z(f)$ is the input impedance in Fig. 3.9, Z_{c1}, L_1, Z_{c2}, L_2, Z_{c3}, L_3 are the characteristic impedance and length to these three pieces of coaxial lines respectively. According to the impedance transformation Eq. (2.13) of transmission line in Chap. 2, we can get

$$Z_{in1} = Z_{c1}\frac{Z_l + jZ_{c1}\tan\beta L_1}{Z_{c1} + jZ_l\tan\beta L_1} \tag{3.8}$$

$$Z_{in2} = Z_{c2}\frac{Z_{in1} + jZ_{c2}\tan\beta L_2}{Z_{c2} + jZ_{in1}\tan\beta L_2} \tag{3.9}$$

$$Z_{in3} = Z_{c3}\frac{Z_{in2} + jZ_{c2}\tan\beta L_3}{Z_{c3} + jZ_{in2}\tan\beta L_3} \tag{3.10}$$

From these equations above, we can get

$$Z_{in3} = \frac{f(Z_{c1}, L_1, Z_{c2}, L_2, Z_{c3}, L_3)}{g(Z_{c1}, L_1, Z_{c2}, L_2, Z_{c3}, L_3)} \tag{3.11}$$

So, Z_{in3} is a function of these variables of Z_{c1}, L_1, Z_{c2}, L_2, Z_{c3} and L_3. To the matched impedance, Z_{in3}, the input impedance should be

$$Z_{in3} = 50 \tag{3.12}$$

Here we need impedance match at three frequency points, there should be three equations of (3.11), in which the implied parameters Z_l and β are related to the frequency.

$$\beta = 2\pi/\lambda_g \tag{3.13}$$

In Eq. (3.8), Z_l is the input impedance at antenna port, and we can get the input impedances from Fig. 3.9 as

$$\begin{cases} Z_{l1} = Z(f = 5.1\,\text{GHz}) = 100 + j25 \\ Z_{l2} = Z(f = 5.5\,\text{GHz}) = 71 - j30 \\ Z_{l3} = Z(f = 5.9\,\text{GHz}) = 34 - j26 \end{cases} \tag{3.14}$$

Equation (3.11) is complex, the real part should be same as the imaginary part for impedance match. So, there are 6 unrelated equations and 6 unknowns to solve.

In fact, the equation here does not need to be solved in a strict sense, because for the general antenna design, it only needs to satisfy the $S_{11} < -10dB$ in band, and it does not need to match completely. Due to the existence of matching margin, the matching equation that should be satisfied eventually changes from (3.12) to

$$Z_{in3(f)} = 50 \pm \Delta Z_{re} \pm j\Delta Z_{im} \tag{3.15}$$

Here, ΔZ_{re} and ΔZ_{im} only are in some range to satisfy the VSWR < 2 in band.

In principle, the matching design proposed here is to achieve impedance matching within a broadband, so in order to have a better matching effect, several more frequency points should be taken to make each frequency point meet the matching Eq. (3.15). However, for each additional frequency point, a complex equation, or two equations, will be added. If we take five frequency points to satisfy the matching equation, then there are 10 equations in total. Generally speaking, these 10 equations are not related. However, the coaxial transmission lines used to design the matching are still three sections as shown in Fig. 3.10, and the total number of unknowns is 6. In fact, 10 equation systems are constructed to solve these 6 unknowns, which leads to overdetermined equations and have no solutions. In order to make the equations have solutions, it is necessary to increase the number of unknowns, that is to say, to increase the number of nodes matching the coaxial transmission line, so that the equations are positive definite or under definite, then there may be solutions to meet the matching equation. But in this way, it will greatly increase the difficulty of solving the equation first, and then increase the number of coaxial transmission lines, which will make the actual structure of the antenna more complex, not only increase the size, but also increase the difficulty of processing, and reduce the practicability of the antenna.

For a system of equations with six unknowns, no matter how many equations there are, it is unrealistic to solve its analytical solution. Here, we choose to get the numerical solution by optimization with the help of simulation software to complete the antenna matching. Here the full wave simulation software HFSS or CST is not used for optimization, the reasons are followed. Firstly, these two full wave softwares are on the base of the field computation algorithms, and they are relatively slow in simulation. Secondly, for this specific task of antenna matching, the radiation part of the antenna has been determined, only the structure of these three pieces of coaxial transmission lines needs to be optimized. Although only the matching part is optimized each time, the whole antenna has to be simulated, which greatly reduces the efficiency of antenna design.

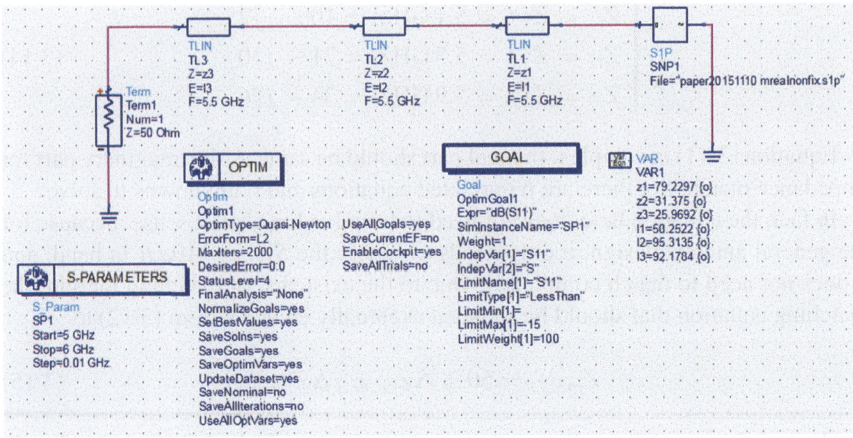

Fig. 3.11 Optimized the impedance matching. Figure reproduced with permission from: [11], © 2015 IEEE

Because of the above two reasons, ADS is chosen as the design tool of matching circuit. Firstly, ADS is based on the circuit method, and the calculation speed is fast. Secondly, the port impedance and return loss of the antenna have been obtained in Figs. 3.8 and 3.9, these curves can be exported as a SIP file, and can replace the radiation part of the actual antenna in circuit simulation. In the simulation, only the matching structure part is calculated, and the optimization interface is shown in Fig. 3.11. Although circuit simulation greatly improves the design efficiency of antenna matching part, there is a problem that it adopts the circuit method, which is fast in calculation but the model is not the actual coaxial transmission line structure. Therefore, after the final optimization result is obtained, the ideal transmission line of ADS needs to be transformed into the actual coaxial structure in HFSS for fine tuning and optimization.

As shown in Fig. 3.12, the results of matching optimization in ADS are compared with the results of S_{11} without matching. The red curve is the results of S11 after matching, and the blue curve is the results of S_{11} without matching. It can be seen that the optimization effect is very obvious. After matching, it shows $S_{11} < -10\,\text{dB}$ in the band of 5.1–5.9 GHz.

The characteristic impedance and electric length of these three transmission lines used for matching are obtained by circuit optimization. According to the electric length and characteristic impedance formula (2.9)–(2.11) in the basic theory of transmission line in Chap. 2, the parameters of ideal transmission line are transformed into the actual coaxial transmission line structure, and the simulation verification and matching optimization fine tuning are carried out in the full wave electromagnetic simulation software such as HFSS. The final optimized result is shown in Fig. 3.13. The comparison between S11 result of feed port after antenna matching and S11 result without antenna matching are shown in Fig. 3.13. The blue curve is the result without antenna matching, and the red curve is the result after antenna matching.

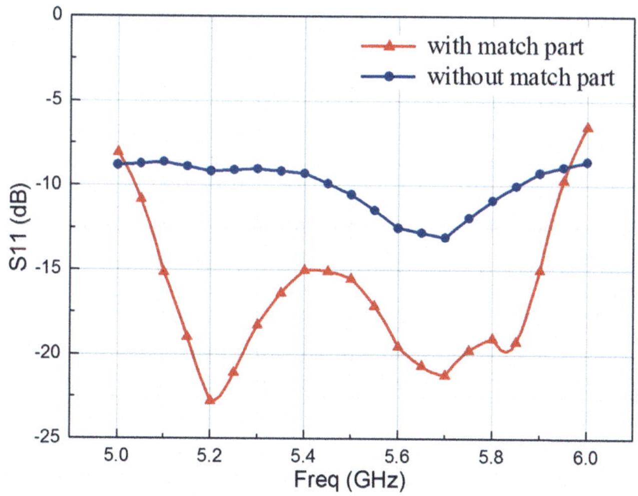

Fig. 3.12 The optimized S-parameter results of antenna without match part compared with antenna adding match part by ADS with ideal transmission line. Figure reproduced with permission from: [11], © 2015 IEEE

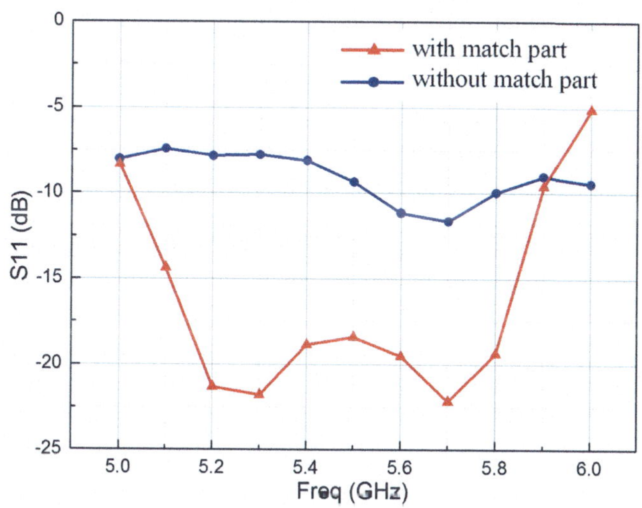

Fig. 3.13 The optimized S-parameter results of antenna without match part compared with antenna adding match part by HFSS with coaxial transmission line

3.3 Antenna Performance Simulation and Verification

Here, CST Microwave Studio software is used to simulate and optimize the omnidirectional CP slots antenna, which is based on the finite integration technique (FIT) method. The final optimized structure parameters' values are listed in Table 3.1.

The magnetic field distribution pictures of the antenna at time $t = 0$, $T/4$, $T/2$, $3T/4$ are given in Fig. 3.14. The H-fields details on A–A section and B–B section of the slots array antenna are given on the left and right separately to show the antenna working principle.

From four status of the magnetic field distribution on the B–B section in a period time (T), the magnetic field are symmetrically distributed and run out from the four slots pairs around the coaxial cylinder shell, and simultaneously varies with time. Although the out magnetic field are not uniform, but the magnetic field in other parallel cut plane along the axis of the cylinder is almost same and periodically deflects with the leaned slot, so that the whole radiation is omnidirectional.

The magnetic field distributions on A–A section of the omnidirectional slots antenna at different times are shown in Fig. 3.14 too. It noted that the radiated magnetic fields from different slots along the axis of the coaxial cylinder are same phase waves in the period, so that the slots array pattern is stable in azimuth plane with high gain in the work band.

In Fig. 3.14a, at $t = 0$, the magnetic fields is strong in the slots close to the feed port side, but weak in the slots pairs near the terminal side, and the fields radiated from the slots pairs is 45° linear polarized wave.

The magnetic fields distribution varying with time, at $t = T/4$, the magnetic fields in the slots pairs near the feed port side become weak, but the magnetic fields changes to peak amplitude and outward direction in the slot close to the terminal side of the slot pairs, which is shown in Fig. 3.14b. And the current fields radiated from the slots pairs is −45° linear polarized wave. At $t = T/2$, the magnetic fields distribution is

Table 3.1 Optimized geometric parameters for the omnidirectional CP antenna

Parameter	Value	Parameter	Value (mm)
Do	32 mm	D4	10
Di	6 mm	L1	6.1
L	161.1 mm	L2	15.8
Wg1	39.59 mm	L3	13.2
Wg2	42.27 mm	L4	40
Wg3	35.77 mm	Ls	23.7
Hg	14 mm	Ws	4.5
D1	4.2 mm	H	11.8
D2	8.7 mm	Dg	3.6
D3	7.6 mm	ds	1.65
β	45°		

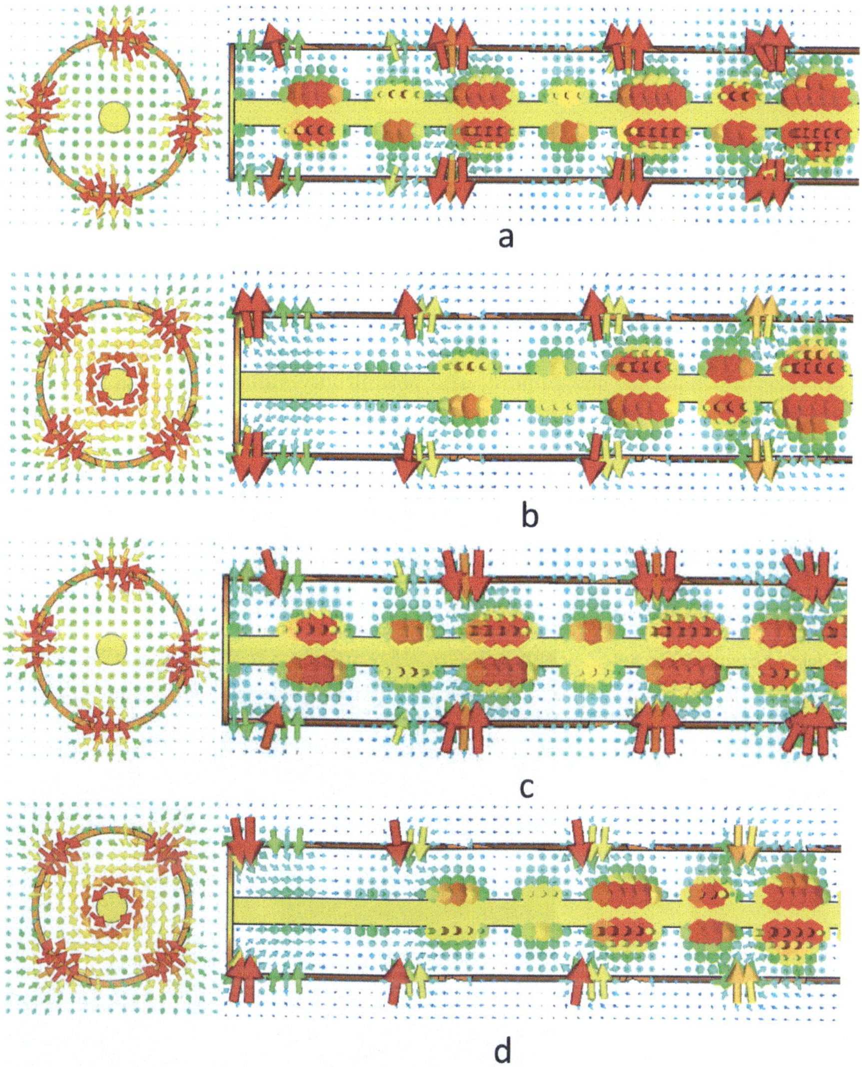

Fig. 3.14 Magnetic fields distribution in A–A section (right) and B–B section (left) of the antenna. **a** t = 0, **b** t = T/4, **c** t = T/2, **d** t = 3T/4. Figure reproduced with permission from: [11], © 2015 IEEE

similar as that at $t = 0$ but with opposite direction. At $t = 3T/4$, the magnetic fields distribution is similar as that at $t = T/4$ but with opposite direction too. The details are shown in Fig. 3.14c, d, respectively.

It is clear that the 45° linear polarized wave and the −45° linear polarized wave exchange with 90° phase difference in all of these slot pairs in the outer shell of the coaxial cylinder.

3.4 Experiment Results Analysis

In this section, the measured results of the designed omnidirectional slots antenna (OSA) are given and compared with the simulation results. The OSA is manufactured by red cooper and Teflon, which is measured in the chamber, and the test scene to the OSA is shown in Fig. 3.15.

The measured return loss and the simulated result are compared in Fig. 3.16. The results showed that the measured impendence bandwidth is 16.4% ranging from 5.05 to 5.95 GHz with $S_{11} < -10$ dB.

The experiment S11 curve is similar as the simulated result with the same trend, but some difference at the high frequency. Here the real Teflon medium to manufacturing is engineering material, and impure. The permittivity value of the manufacturing material may be smaller than the Teflon in the material library in CST in simulated value of $\varepsilon_r = 2.1$.

The average axial ratio (AR) and gain of the measured and the simulated results in azimuth plane (*xoy*-plane) are compared in Fig. 3.17. The results showed that the measured gain is 5–7 dBic during the work band 5.1–5.9 GHz, and the corresponding average AR below 3 dB is in the same band 5.1–5.9 GHz. In Fig. 3.17, the designed

Fig. 3.15 Photograph of the omnidirectional CP antenna being tested in the anechoic chamber. Figure reproduced with permission from: [11], © 2015 IEEE

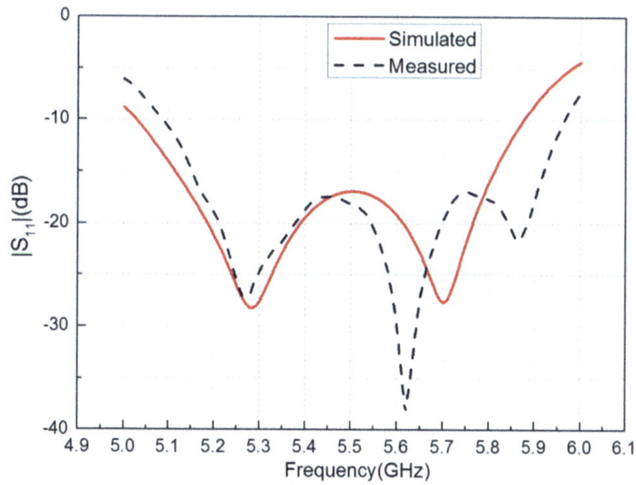

Fig. 3.16 Measured and simulated S_{11} of the omnidirectional CP antenna. Figure reproduced with permission from: [11], © 2015 IEEE

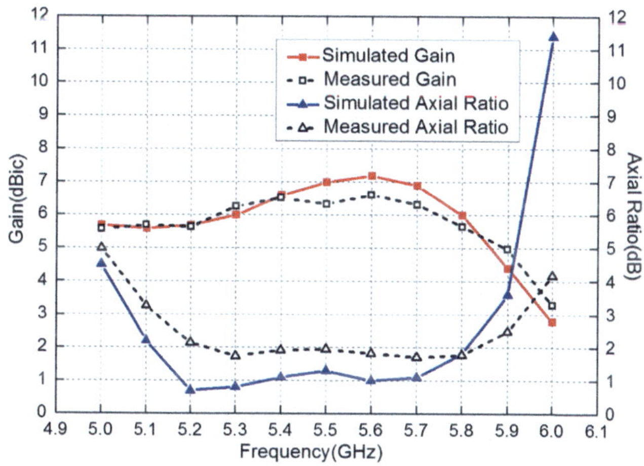

Fig. 3.17 Measured and simulated gain and average axial ratio results of the omnidirectional CP slot antenna. Figure reproduced with permission from: [11], © 2015 IEEE

OSA can work in omnidirectional radiation with stable high gain and good CP performances in the work band. Here, the measured gain is 0.8 dB lower than the simulated result near the center frequency, and there are some degradations between the measured and simulated average AR curves too.

Besides the impure medium of the Teflon, measuring process error is another reason to the difference between the measured and the simulation too. Thirdly, the main reason should be the machine error. The designed OSA is complicated, and it is sensitive to the slots pairs sizes and relative positions. For example, the slot is rectangular, which is constructed by cutting a cubic from the cylinder. But it is realized by a circle milling cutter around the axis in the manufacturing, so that the slot deformed a little, and the angle of the slot is not 90° but small arc.

The measured and the simulated RHCP and LHCP normalized radiation patterns in the azimuth plane (*xoy*-plane) and elevation plane (*yoz*-plane) are compared in Fig. 3.18a, b and c at 5.2 GHz, 5.5 GHz and 5.8 GHz, respectively. The measured results are consistent with the simulation results with little divergences, and the cross-polarization is less 16 dB than the co-polarization beam in the azimuth plane. All these performances again certificate the good omnidirectional CP characteristics of the OSA.

3.5 Improved Omnidirectional CP Coaxial Slots Antenna with Compact Size and High Gain

In the above section, the design of omni-directional circular polarization antenna based on coaxial slot array is given, but the gain is limited. In this section, how to design an omni-directional circular polarization antenna with higher gain is discussed in detail. Because the antenna is based on the slot array to achieve high gain, in order to further improve the antenna gain, the direct choice is to increase the number of slots, that is, increase the number of array elements to design an omnidirectional circularly polarized antenna with higher gain.

In addition to the research on improving the gain, as mentioned in the previous section, because the structure of the antenna is simple and completely symmetrical, only the feed direction needs to be changed to change the polarization rotation direction of the antenna, that is, the antenna is actually a circularly polarized antenna with controllable polarization. The antenna design in this section is also a confirmation of this conclusion. In the last section, a left-handed circularly polarized antenna is designed by changing the feeding direction of the antenna.

3.5.1 Improved Antenna Geometry Structure

The improved omnidirectional CP slots antenna is proposed in Fig. 3.19. It is based on the OSA in above section. The two orthogonal slots pair is the basic radiation unit, and there are four slot pairs uniformly distributed in the outer conductor shell around the coaxial cylinder axis. Same as the original OSA in above section, the slot leans an angle of 45° or −45°, and both slots in a pair are arranged $\lambda_g/4$ (λ_g is the medium wavelength) distance away each other along the axis, which ensure the leaked waves from the slot pair is orthogonal and with 90° phase difference each other. And the

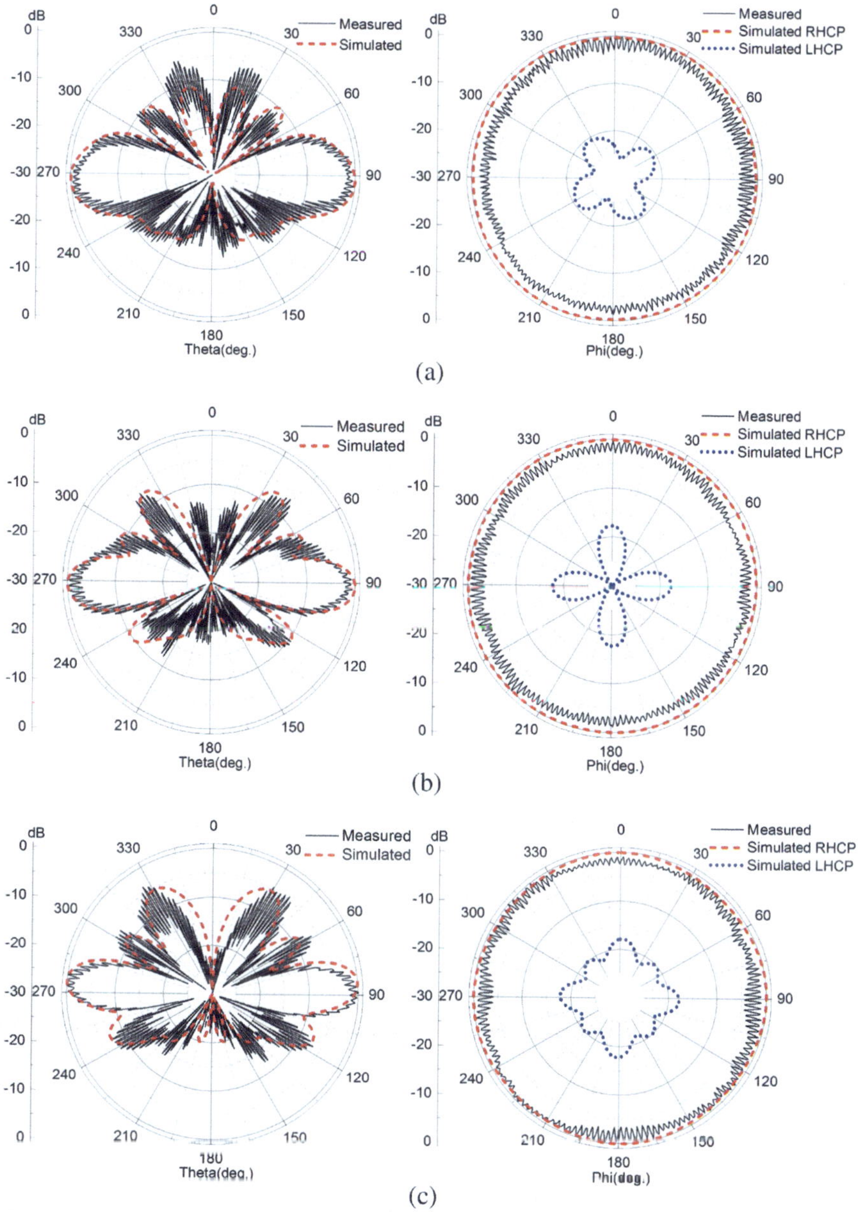

Fig. 3.18 Simulated RHCP and LHCP normalized radiation patterns and measured normalized radiation pattern in the E plane (yz-plane on the left) and the omnidirectional plane (xy-plane on the right). **a** 5.2 GHz. **b** 5.5 GHz. **c** 5.8 GHz. Figure reproduced with permission from: [11], © 2015 IEEE

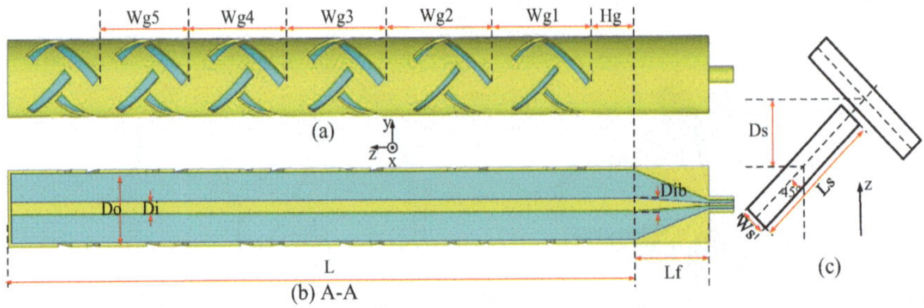

Fig. 3.19 The Geometry structure of the high gain omnidirectional CP antenna, **a** side view, **b** A–A section view, **c** geometry structure of the perpendicular slot pairs

designed improved circular polarized OSA performs left hand circularly polarized property finally.

In further, six rounds of the basic radiation units along the coaxial cylinder axis are introduced to construct slots array to get higher gain. In order to ensure the waves radiated from every element of the slot array with the same phases, we arrange these elements at an interval of λ_g with each other. We choose Teflon with a permittivity $\varepsilon_r = 2.1$ as medium, so the interval between the slot array elements is $\lambda_g = 0.7\lambda_0$ (λ_0 is the wavelength in free space). To achieve wide work band, little differences in interval between the six rounds slot array units are predesigned, which is useful to adjust the admittance distribution and leaked field distribution to realize a relatively wide available impedance bandwidth. These six rounds of basic radiation units are separated into three groups based on their slot size. Two adjacent radiation units is a group, and their size is same. From feed side to the terminal end, Six rounds of slots units are cut more and more short and wide in the outer shell of the coaxial cylinder. In fact, the radiation beam will diverge from the azimuth plane and the gain decreases if all of these slots are designed with same size in the outer shell of the cylinder. So, the little differences in the slots array can compensate the defects of the same size omnidirectional CP slot array. The simulation results prove the biggest gain beam with omnidirectional radiation in the azimuth plane without deviation. Similar as the original OSA, the coaxial tapered coaxial cylinder structure is introduced to connect the radiation slots array with large size and small SMA adaptor, which is useful to wide band impedance matching and compact antenna.

Certainly, more slots increase the antenna performances with compact size, which is a good method to extend the design method to this kind OSA.

3.5.2 *Improved Antenna Design*

High gain omnidirectional CP antenna is designed based on the former omnidirectional CP antenna, the omnidirectional radiation and CP performance are similar, and the interval is $\lambda_g/4$ between two adjacent slots, the CP principle is same as shown in Eqs. (3.1)–(3.7).

The feed direction of the antenna designed in this section is shown in Fig. 3.19, which is opposite to the feed direction of the previous section. And the phase of E_x from the slot pair as shown in Fig. 3.3 lags behind the phase E_y, that is $\varphi = -90°$ and $\alpha = -\omega t$. The rotation angle of the electric field vector increases in clockwise direction with time, results in LHCP wave, and the antenna is a LHCP antenna.

The omnidirectional performance of the antenna is guaranteed by a circle of four pairs of slots on the outer wall of the coaxial. According to the analysis in the previous section, this circle of slots serves as an omnidirectional circular polarization unit. In order to get a higher gain, an antenna with six circles of slots to form a six-element array is designed in this section. These six circles slot of the antenna is distributed along the central axis in the feed direction, which is equivalent to the serial feed array antenna. The array element spacing is a medium wavelength λ_g, which ensures that the feed phase is in same phase in these slots, so that the beam direction of the array is perpendicular to the array element arrangement direction, i.e. in the horizontal direction.

The antenna is based on the structure of coaxial waveguide transmission line. Since these six elements in the outer wall of the coaxial line distribute along the axis and fed in serials, the six circle slots successively can be equivalent to the six radiation impedances successively connecting on the transmission line, as shown in Fig. 3.20 [15].

When the antenna is fed from the feed port, the electromagnetic wave passes through these six equivalent radiation impedance networks in turn, in which IL_i represent the insertion loss of the ith equivalent impedance network. The insertion loss is

$$IL = 10\log(P_o/P_i) \tag{3.16}$$

where, P_o and P_i represent network output and input power respectively. According to formula (3.16), the insertion loss of the network IL represents the proportion of the part of the electromagnetic energy radiated out when passing through a circle of slots.

When the slot sizes on the outer wall of the coaxial waveguide transmission line are same, the parameters of the equivalent impedance network of each slot should be same, so the six equivalent impedance networks have the same insertion loss

$$IL_1 = IL_2 = IL_3 = IL_4 = IL_5 = IL_6 \tag{3.17}$$

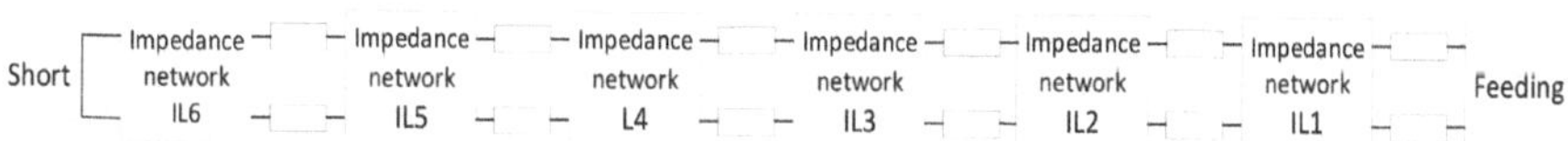

Fig. 3.20 Equivalent transmission line and impendence network of the antenna

Assume P_i is the input power, from Eq. (3.16)

$$P_o = 10^{\frac{IL}{10}} P_i \tag{3.18}$$

If we set $P_i = 1$ and $Q_i = 10^{\frac{IL_i}{10}}$, and assume PW_i is the radiated power from the ith slot respectively, so

$$PW_i = (1 - Q_i)Q_i^{i-1} \tag{3.19}$$

According to Eq. (3.17), the proportion of the energy radiated from the slot to the total energy at the feed port reference plane is same as the equivalent impedance network. For $Q_i < 1$, it can be seen from Eq. (3.19) that in the actual feeding environment, the feeding amplitude of each unit decreases along the axis. According to the theory of antenna array in Chap. 2, when the amplitude of each element is equal, the gain of the antenna is the highest. The decreasing amplitude distribution of the antenna array will reduce the gain of the antenna array and increase the side lobe level of the antenna. For this kind of linear array with decreasing amplitude distribution, in extreme cases, when the feed amplitude difference between the first unit and the last unit is too large, it can be considered that the last unit has almost no gain contribution to the whole array, that is to say, the last unit with too small amplitude can be ignored.

Due to the inherent structure limit of the antenna, the feeding port is on one side of the coaxial transmission line, and six elements are distributed along the axis in turn, which is equivalent to the serial feeding array, so this unequal amplitude feeding effect cannot be completely eliminated. In order to prevent the last two laps of the antenna farthest from the port from being ignored due to this effect, the slot size is adjusted to compensate and balance this unequal amplitude feeding effect. By changing the slot size of coaxial transmission line, the equivalent impedance network is changed to decrease the insertion loss of equivalent impedance network closer to the feeding port, that is to say, the proportion of energy radiated smaller from the slot. In terms of absolute value, the energy radiated by the slot closer to the feed port is still more, but this does balance the feed amplitude of the array element, so that the final array element will not be ignored because the feed amplitude is too small to contribute to the array gain.

The specific method in this section is to divide the six circles slot into three groups. Each group has the same slot size, but the slot sizes between the two groups are different from each other. The specific size values are shown in Table 3.2. In this way, there are three different insertion loss values for the six radiated impedance networks equivalent to the six-circle gap, and the closer to the feed port network, the smaller the insertion loss is.

$$IL_1 = IL_2 < IL_3 = IL_4 < IL_5 = IL_6 \tag{3.20}$$

So,

Table 3.2 The parameters of the high gain omnidirectional CP antenna

D_o (mm)	28.8	L_{s1} (mm)	24.6
D_i	4.7	W_{s1}	2.3
L	253.1	L_{s2}	23.2
W_{g1}	39.9	W_{s2}	3.6
W_{g2}	42.7	L_{s3}	21.6
W_{g3}	40.1	W_{s3}	3.7
W_{g4}	37.9	D_s	11.6
W_{g5}	36.2	H_g	17.5
L_f	30	D_{ib}	5

$$Q_1 = Q_2 < Q_3 = Q_4 < Q_5 = Q_6 \tag{3.21}$$

From Eq. (3.19), with the increasing of Q_i in turn, the radiated energy difference between the three groups of six circles slot is decreasing. In this way, the equivalent feed amplitude of six array elements is balanced. and reduced the feed amplitude difference between the equivalent feed amplitude of the two farthest circle slot and the equivalent feed amplitude of the nearest two laps.

The magnetic field distribution on the cross section at 5.5 GHz after changing the slot size, i.e. equalizing the equivalent feed antenna model are shown in Fig. 3.21. Here, the magnetic field distribution diagram refers to the absolute value of the magnetic field vector on the antenna section in a period, that is, the magnetic field in the diagram is scalar, and its value is the maximum value of the magnetic field in the whole working period of the antenna. Therefore, from such a maximum distribution diagram of magnetic field, it can be seen that the energy distribution at the six circles slot of the antenna or the equivalent feeding amplitude of six elements decreases with the increase of distance from the feeding port, and the antenna is an in-phase array with decreasing amplitude distribution.

Figure 3.22 shows the simulation model of the six circles slot antenna without changing the slot size of the outer wall of the antenna on the basis of the previous

(a)

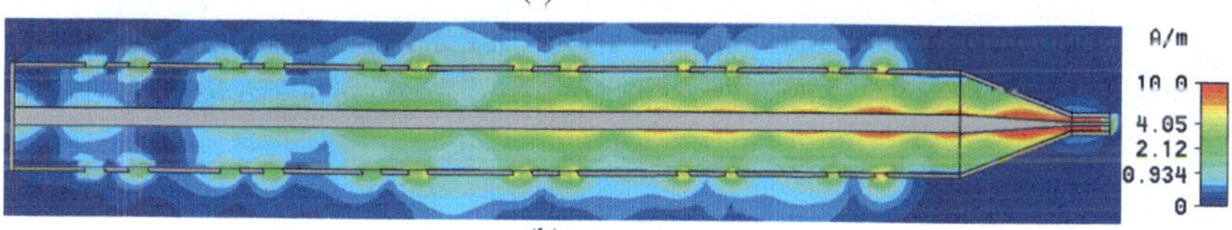

(b)

Fig. 3.21 Magnetic field distribution on A–A section of the antenna with different size slots. **a** Antenna model **b** magnetic field distribution at 5.5 GHz

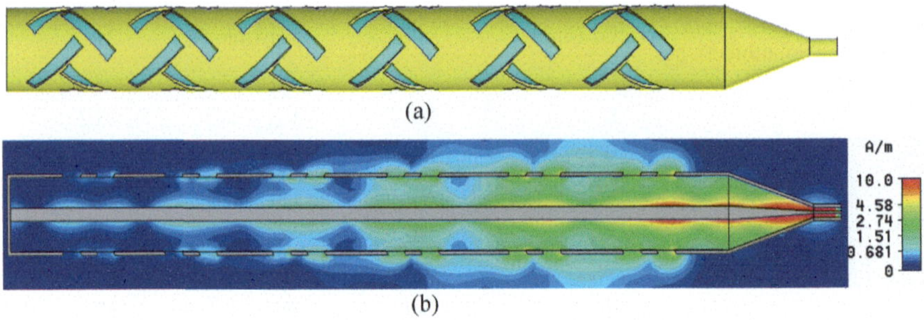

Fig. 3.22 Magnetic field distribution on A-A section of the antenna with different size slots. **a** Antenna model **b** magnetic field distribution at 5.5 GHz

section of the antenna, and the cross-section magnetic field distribution diagram at the frequency of 5.5 GHz. Similarly, the magnetic field here gives the absolute value of the maximum value. In order to better display the contrast effect, the color scale and coordinate scale used for this magnetic field distribution map and the magnetic field distribution map are same as those shown in Fig. 3.22b, and the range from the minimum value to the maximum value of the two maps is $0 \sim 10A/m$. Compared with the magnetic field distribution of the variable size slot antenna shown in Fig. 3.23b, it is obvious that in this case, the gradient of the magnetic field distribution is more obvious, the energy distribution in these six circles slot is much more uninform, and the feed amplitudes of the equivalent array elements are more unbalanced. The energy distribution in the circle slot farthest from the feed port is smaller than that of the nearest circle slot. Therefore, the gain of the equal size slot antenna shown in Fig. 3.22a is lower than that of the unequal size slot antenna shown in Fig. 3.21a. In both cases, the gain pattern of the antenna is shown in Fig. 3.23. It can be seen from the figure that the gain of the antenna with unequal slot size is

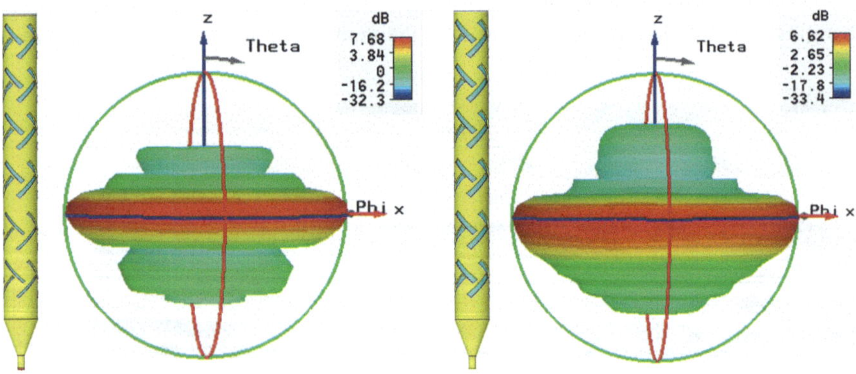

Fig. 3.23 Radiation patterns of the antennas with (**a**) different size slots (left) and (**b**) same size slots (right) at 5.5 GHz

7.68 dBic, and the gain of the antenna with equal slot size is 6.62 dBic. Figure 3.17 shows the gain of the OCPA with four circles of slots on the outer wall. It is found that the gain of the antenna at 5.5 GHz frequency point is about 7 dBic, which is higher than that shown in Fig. 3.23b, the gain of the antenna with six circles of equal size slot on the outer wall at this frequency, indicating that the added two circles of slot have no contribution to the array. It is necessary to adjust the size of the slot, so that the antenna can achieve the purpose of increasing the antenna gain by increasing the number of equivalent elements.

The gain can be increased by increasing the number of apertures, that is to say, increase the number of array elements. But it is not achieved by increasing the number of apertures as long as you want to increase the gain. If you want to increase the antenna gain by increasing the number of slots, for example, adding two circles slot based on the six circles slot antenna in Fig. 3.22a to design an antenna of the same type with eight circles of slot. As analyzed on above, it is no use that just simply adds the last two circles of slot, and the equivalent feed amplitude of the last two circles of slot away from the feed port is too small.

In order to achieve the goal of balanced feeding, it is necessary to change the slot size to change the insertion loss of the equivalent radiation impedance network. However, according to the basic theory of slot antenna in Chap. 2, slot and symmetrical dipole antenna complement each other, and the size of slot is related to its working frequency band. To change the insertion loss of the equivalent radiation impedance network, the slot available frequency must be in the original working frequency range of the antenna design. Otherwise, even if the insertion loss of the equivalent impedance network of the antenna barely conforms to the feed amplitude equalization of the array element (being equivalent from the slot of each circle), the S_{11} result of the antenna will be very poor, and it will not achieve the effect of array formation in the band, and the gain in the band of interest will not be improved, which in turn will affect the axial ratio in the band. Therefore, the size change of slot is constrained by the performance of slot antenna itself, which cannot be changed without limitation.

For the high gain OCPA designed in this section, the short circuit is directly adopted at the end as in the previous section. The difference is that the slot in the last circle remains completely, and the possible total reflection effect of electromagnetic wave caused by the short circuit in the last section is not considered. In the last segment, the reflected wave will deteriorate the circular polarization performance of the antenna and the axial ratio will become larger. So, there is a tangent angle in the last circle of the antenna in the last segment to reduce the reflection. This method can protect the circular polarization performance of the antenna, but it will affect the radiation efficiency and the gain of the antenna because of the slot cutting in the last circle. The purpose of this section is to design an antenna with higher gain, and the magnetic field distribution of the section is shown in Figs. 3.21 and 3.23. The electromagnetic energy is very small in the last circle of the slot, and the reflected wave energy will be smaller after the electromagnetic wave radiates outward through the last circle of the slot, which will cause some deterioration of the circular polarization performance of the antenna, but the antenna gain is ensured by sacrificing a little circular polarization performance of the antenna. Here, as in the previous section, we do not adopt the last circle slot angle cutting design at the end, and choose to ignore the influence of the axial ratio deterioration caused by the reflection from the short circuit. The final simulation results are given in the next section, we can see that the circular polarization performance of the antenna is still very good.

The slots size of antenna in this section are different, and each slot resonant frequency corresponding to the equivalent magnon antenna is different, so the composed array has better return loss in thc band. Comparing with the previous section of the same size four circle slot antenna, it is not necessary to specially design multiple transmission lines for matching design. Here, a coaxial gradient transmission line is directly used to match the large-scale coaxial line of the antenna to the standard SMA connector. As shown in Fig. 3.19, the length of the coaxial gradient line L_f and the diameter D_{ib} of the inner conductor at the plane joint the antenna are designed and selected to adjust the return loss S_{11} of the antenna. Scanning the inner conductor diameter D_{ib} of the coaxial gradient line, as shown in Fig. 3.24, the return loss of the antenna in the band becomes better with the increase of the inner conductor diameter D_{ib}. The results of scanning L_f are shown in Fig. 3.25. The length has little effect on S_{11} in the band, but it will have an effect on the matching of the antenna at the side frequency of 5.1 GHz. According to the scanning results, we finally select the parameters $L_f = 30\,\text{mm}$ and $D_{ib} = 5\,\text{mm}$ to ensure $S_{11} < -10\,\text{dB}$ in the frequency band of 5.1 GHz–5.9 GHz. The final structural parameters of the designed antenna are listed in Table 3.2.

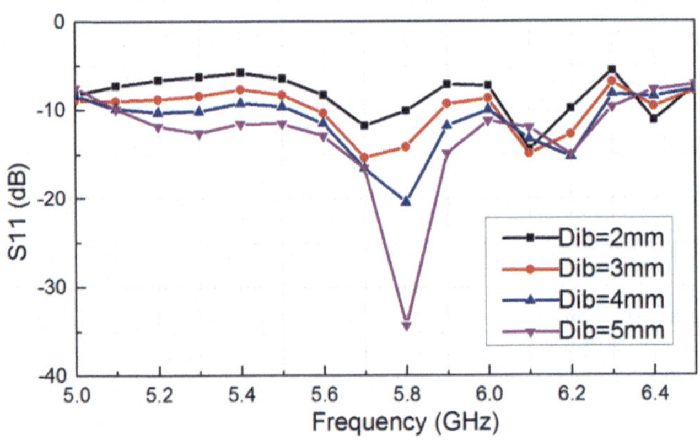

Fig. 3.24 Effect of the diameter of the inner conductor of the coaxial tapered line for S11 results of the antenna

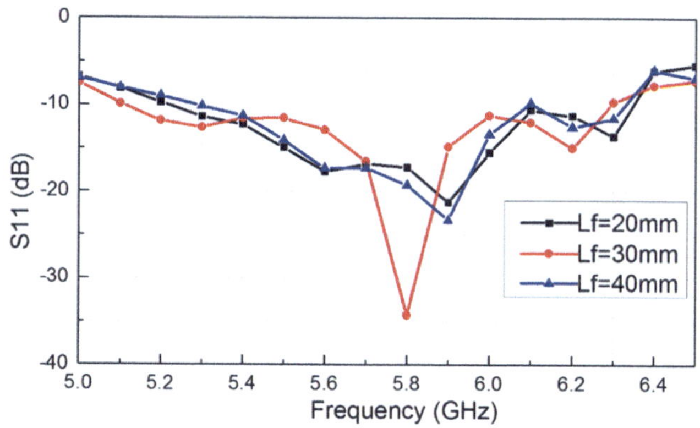

Fig. 3.25 Effect of the length of the coaxial tapered line for S11 results of the antenna

3.5.3 Antenna Performance Discussion

The simulated return loss result is presented in Fig. 3.26. It can be seen that the available bandwidth with $S_{11} < -10$ dB is about 21% over 5.1–6.3 GHz. The realized gain and the average AR in the azimuth plane (*xoy*-plane) is given in Fig. 3.27. It showed that the realized gain is about 6.5–9 dBic during 5.1–5.9 GHz, and the average AR below 3 dB covers the band 5.1–5.9 GHz too. It is clear that the results in Figs. 3.26 and 3.27 show omnidirectional CP bandwidth is 14.55% with a stable high gain in the wide impedance bandwidth.

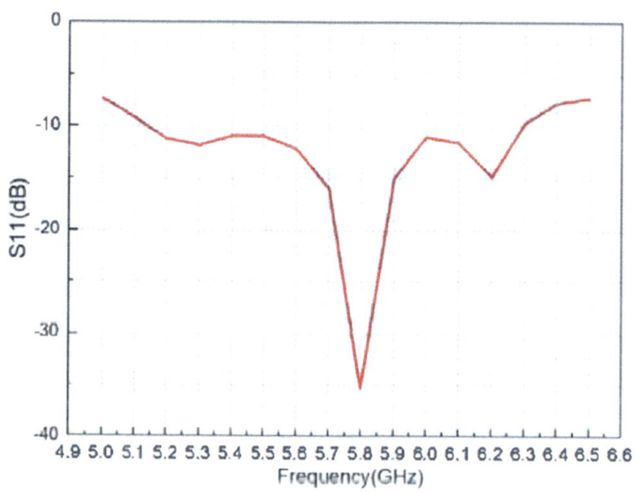

Fig. 3.26 S_{11} of the omnidirectional CP antenna. Figure reproduced with permission from: [14], © 2015 IEEE

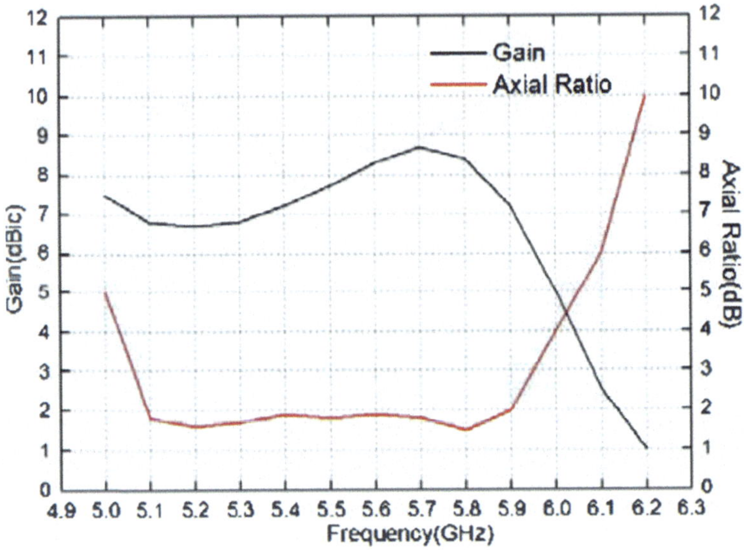

Fig. 3.27 The simulated gain and average axial ratio in the H plane (xy-plane). Figure reproduced with permission from: [14], © 2015 IEEE

The 2D axial ratio curves at 5.5 GHz in the E-plane and H-plane are shown in Fig. 3.28. The AR curve in E-plane clearly shows that the beam width with AR < 3 dB is about 20 degrees in elevator plane. The AR curve in H-plane shows that

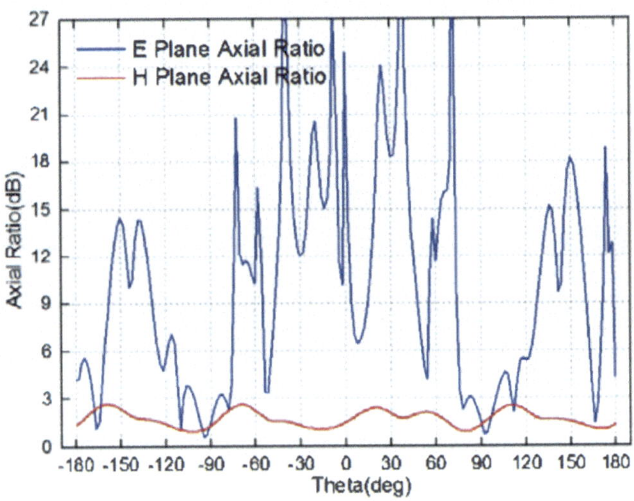

Fig. 3.28 E-plane and H-plane axial radiation pattern at 5.5 GHz. Figure reproduced with permission from: [14], © 2015 IEEE

axial ratio is 0.8 dB–2.9 dB, which means very good CP property is realized in the azimuth plane. The simulated RHCP and LHCP far fields in the azimuth plane (H-plane, or *xoy*-plane) and E plane (*yoz*-plane) at 5.1 GHz, 5.5 GHz and 5.9 GHz are shown in Fig. 3.29, respectively. It is clear that the LHCP radiation is main part, and the cross-polarization, RHCP beam in the *xoy* plane is less 15 dB than the co-polarization, LHCP beam. The co-polarization far fields in xoy plane are almost a standard circles, and show good omnidirectional radiation performances. The far fields curves in E-plane of the LHCP pattern shows good symmetrical character, and the HPBW (half power beam width, or 3 dB beam width) is about 18 degrees.

3.6 Summary

The omnidirectional CP antenna by cutting slot array on the outer conductor shell of the coaxial cylinder was designed by leaking the traveling wave in coaxial wire. It performs stable omnidirectional CP radiation with high gain in wide bandwidth. The designed antenna is manufactured, and the experiment results are almost consistent with the simulation. The measured results show that the available bandwidth with $S_{11} < -10$ dB and AR $<$ 3 dB is about 15.4% over 5.1–5.9 GHz. And the measured gain is stable and being larger than 5 dBi in the work band. The antenna radiated good omnidirectional CP far fields with cross-polarization less 16 dB than the co-polarization main beam in azimuth plane. It is obvious that the proposed OSA has good potential to strengthen the wireless communication ability.

In further, it is available to design the OSA with expected performances by adjusting the number of the basic slots units to satisfy the different demands.

An improved OSA with high gain and CP character based on the traveling wave in the coaxial cylinder structure and leaking from the slot arrays in the outer conductor shell is designed. The improved OSA can work in 5.1–6.3 GHz, about impedance bandwidth of 21.05% with $S_{11} < -10$ dB. The realized gain of the improved OSA is larger than 6.5dBi in the work band, and the corresponding AR is less than 3 dB in the azimuth plane. So the improved OSA is a good omnidirectional CP antenna, which has a good promotion in wireless communication.

Fig. 3.29 LHCP and RHCP gain radiation patterns in the xy plane and yz plane at (**a**) 5.1 GHz (**b**) 5.5 GHz (**c**) 5.9 GHz. Figure reproduced with permission from: [14], © 2015 IEEE

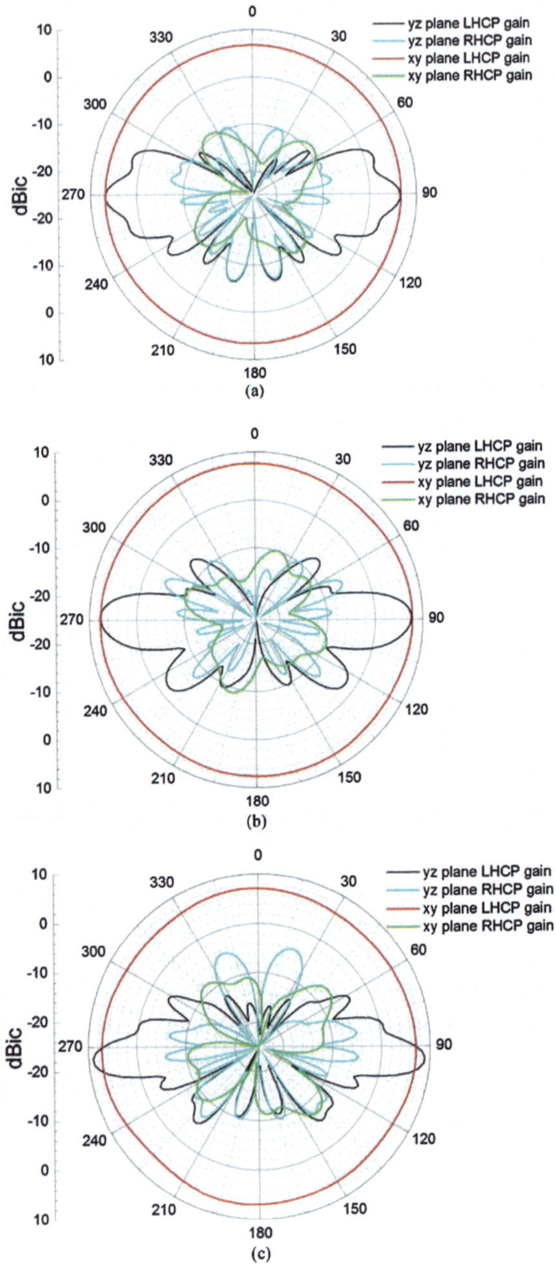

References

1. K. Sakaguchi, N. Hasebe, A circularly polarized omnidirectional antenna, in *Proc. 8th Int. Conf. Antennas Propag. (ICAP'93)* (1993), pp. 477–480
2. B.C. Park, J.H. Lee, Dual-band omnidirectional circularly polarized antenna using zeroth- and first-order modes. IEEE Antennas Wireless Propag. Lett. **11**, 407–410 (2012)
3. A. Narbudowicz, X.L. Bao, M.J. Ammann, Dual-band omnidirectional circularly polarized antenna. IEEE Trans. Antennas Propag. **61**(1), 77–83 (2013)
4. W.Q. Cao, A.J. Liu, B.N. Zhang, Dual-band spiral patch-slot antenna with omnidirectional CP and unidirectional CP properties. IEEE Trans. Antennas Propag. **61**(4), 2286–2289 (2013)
5. Y.M. Pan, K.W. Leung, K. Lu, Omnidirectional linearly and circularly polarized rectangular dielectric resonator antennas. IEEE Trans. Antennas Propag. **60**(2), 751–759 (2012)
6. W.W. Li, K.W. Leung, Omnidirectional circularly polarized dielectric resonator antenna with top-loaded Alford loop for pattern diversity design. IEEE Trans. Antennas Propag. **61**(8), 4246–4256 (2013)
7. B. Li, S. Liao, Q. Xue, Omnidirectional circularly polarized antenna combining monopole and loop radiators. IEEE Antennas Wirel. Propag. Lett. **12**, 607–610 (2013)
8. Y. Yu, Z. Shen, S. He, Compact omnidirectional antenna of circular polarization. IEEE Antennas Wirel. Propag. Lett. **11**, 1466–1469 (2012)
9. X. Quan, R. Li, M.M. Tentzeris, A broadband omnidirectional circularly polarized antenna. IEEE Trans. Antennas Propag. **61**(5), 2363–2370 (2013)
10. B. Zhou, High gain omnidirectional circularly polarized antenna with slots on coaxial waveguide, Mater thesis, Shanghai Jiao Tong University (2016)
11. B. Zhou, J. Geng, X. Bai, L. Duan, X. Liang, R. Jin, An omnidirectional circularly polarized slot array antenna with high gain in a wide bandwidth. IEEE Antennas Wirel. Propag. Lett. **14**, 666–669 (2015). https://doi.org/10.1109/LAWP.2014.2376961
12. K. Iigusa, T. Teshirogi, M. Fujita, S.-I. Yamamoto, T. Ikegami, A slot-array antenna on a coaxial cylinder with a circularly polarized conical beam. Electron. Comm. Jpn. Pt. I **83**, 74–87 (2000)
13. M.E. Bialkowski, P.W. Davis, A linearly polarized radial-line slot-array antenna with a broadened beam. Microwave Opt. Technol. Lett. **27**(2), 98–101 (2000)
14. B. Zhou, J. Geng, R. Jin, X. Liang, A Compact omnidirectional CP coaxial sslots antenna, in IEEE Mape, Shanghai, China, 10.28-30 (2015)
15. W. Wu, C. Liang, Microwave network and its application [M]. National Defense Industry Press (1980)

Chapter 4
Principles of Omnidirectional Dual-CP Slot Array Antenna

4.1 Introduction

In wireless communication, there will be any polarization electromagnetic waves, including the intentionally loaded wave or depolarized wave by the environment, in which the antenna with dual circular polarization performances, left hand circular polarization (LHCP) or right hand circular polarization (RHCP), is very applicable and valuable in the complicated EM environment. Dual circular polarization omnidirectional antenna has been widely used in wireless transmission, radio broad casting and navigation etc. [1–3]. Especially, the compact omnidirectional CP antenna has been the key and bottle net to the wireless system. And there has been some results reported recently. A miniaturization of omnidirectional CP antenna by folding the antennas patch underneath itself to decrease size has been given in [4]. The antenna is small, but the impedance bandwidth is very narrow, only 0.6% (2.392–2.407 GHz), and the CP performance is poor with AR > 3 dB in some directions in the azimuth plane. In [5], a CP antenna is designed by pasting the inclined slits to the sidewalls of the rectangular dielectric resonator (RDR), and a rectangular part is cut from the top wall of the linearly polarized (LP) rectangular dielectric resonator antenna (RDRA). The excited degeneracy modes combined into the CP fields. But the available impedance bandwidth is narrow, and the out-of-roundness of the pattern in the azimuth plane is larger than 5 dB, so that the omnidirectional performance is poor. A dual circular polarization antenna is shown in [6], which is fed by hybrid coupler with 90° phase shift to combine the CP wave. Here the CP wave works at the second-order mode, and the pattern is up warping and becomes conical beam. In [7], a dual-band omnidirectional micro-strip antenna with dual circularly polarized character is presented. But the antenna only works in single RHCP and LHCP mode in different bandwidths respectively, or it is a single CP antenna in the given frequency band. Literature [8] presents an omnidirectional antenna with dual-CP performance, and 4 tilted dipoles with parasitic elements are introduced to generate each sense of CP.

J. Geng, *Omnidirectional Slots Antenna*,
https://doi.org/10.1007/978-981-15-9089-4_4

In this chapter, the principle of an omnidirectional coaxial slot array antenna with dual CP character by cutting perpendicular slot pairs in the outer conductor shell of the coaxial cylinder around and along the axis is proposed [9]. The radiation part is the same slots array in the outer conduct shell for RHCP and LHCP wave, but the feed ports for the RHCP and LHCP wave are different, and they are designed in the two side of the coaxial cylinder respectively. The designed omnidirectional slots array antenna with dual CP character can work in a wide bandwidth, and it radiates good omnidirectional patterns with small axial ratio in RHCP or LHCP. The structure of the omnidirectional slots array antenna with dual CP character and the design principle are given in Sect. 4.2. The detailed parameters discussion and the experiment results are shown in Sect. 4.3. The summary of this chapter is given in the end.

4.2 Geometry Structure, Principle and Antenna Design

4.2.1 Antenna Structure

The presented omnidirectional antenna with dual CP character in this paper is shown in Fig. 4.1. The idea is from the principle of the omnidirectional slots antenna (OSA) with single CP performance in [12, 13]. And the antenna is still constructed with slots array in the shell of the coaxial cylinder structure. Two ports corresponding to left hand circularly polarized (LHCP) and right hand circularly polarized (RHCP) are assigned in the two side of the coaxial cylinder, respectively. The original coaxial cylinder in fact is a thick coaxial wire with characteristic impedance 50 Ω in the EM model, which is same as the standard SMA adaptor with characteristic impedance 50 Ω. Similar as the original OSA model in Chap. 3, there are two coaxial tapered wire to connect the radiation part with slots array and the SMA adaptors in two side of the cylinder, and they are same to achieve the port impedance matching.

Here, the leaned slots in the outer conductor shell is still the radiation element. Two adjacent orthogonal slots combine a pair, and four slots pair uniformly distribute around the coaxial cylinder to construct the radiation unit. And four rounds of radiation slots units are arranged along the axis of the coaxial cylinder.

The basic slot pair structure including two perpendicular slots is given in Fig. 4.2. One slot leans 45° angle from the axis of the coaxial cylinder, and another slot leans −45° angle from the axis of the coaxial cylinder, and the interval distance between these two orthogonal slots is $\lambda_g/4$ (λ_g is the wavelength in medium) along the feed direction. When the electric fields radiated from these two perpendicular slots, the fields are orthogonal and including a 90° phase difference with each other, So, the fields from the two orthogonal slots are combined into CP wave in the space. Since the distance between the adjacent slots is $\lambda_g/4$, the reflected EM field from radiation slots back to the feeding point is just cancelled by the incident EM signal.

Table 4.1 Optimized geometric parameters of the omnidirectional CP antenna

L	153.5 mm	Wg	40 mm
Di	7.6 mm	Do	28 mm
Ls	24.4 mm	Ws	4.3 mm
Ds	11.8 mm	Lf	18 mm

In theory, the slot can be regarded as magnetic current, of which the length of the slot is $\lambda_g/2$ in the fed direction. Four slots pairs are uniformly distributed in the outer conductor shell around the coaxial cylinder to realize the omnidirectional radiation. Four rounds of slot pairs are cut in the outer conductor shell along the axis of the coaxial cylinder to construct the slots array to achieve higher gain by the four identical rounds of slots pairs. To get omnidirectional radiation beam in the azimuth plane with higher gain, the radiated wave from every round of slots pairs should transmit in same phases, which means the distance between the neighbored rounds of slot pairs is λ_g to realize same phase to the radiated wave.

The outside form of the antenna is a uniform cylinder. And the thickness of the outer conductor shell is $t = 1.3$ mm. And the Teflon medium with a permittivity of $\varepsilon_r = 2.1$ is used to fill the ring space between the outer conductor shell and the inner conductor cylinder. The design process to the proposed antenna is carried out in CST and the optimized parameter values are listed in Table 4.1.

4.2.2 *Dual CP Principle*

Two adjacent slots model is given in Fig. 4.2, which is cut in the outer conductor shell, and they are perpendicular with each other. The reference coordinate system is created along the slot pairs, in which Y and X axis are just the directions with the polarization of the E-fields transmitted by the two slots individually, and the propagation direction of the electric fields is just the +Z axis [10, 11]. Assume the total electric fields being $\vec{E}$, and $\vec{E}_1$ and $\vec{E}_2$ are the fields radiated from one slot and another slot respectively. Then they can be written as

$$\vec{E} = \vec{E}_1 + \vec{E}_2 = E_x\vec{a}_x + E_y\vec{a}_y \tag{4.1}$$

Since the distance between the two slots is $\lambda_g/4$, E_{x0} and E_{y0} ($E_{x0} > 0$, $E_{y0} > 0$) are supposed being the amplitudes of the E_x and E_y, and E_y is laggard to E_x with a phase difference of φ. Then (4.1) can be rewritten as

$$\vec{E}(z, t) = E_{x0}\cos(\omega t - kz)\vec{a}_x + E_{y0}\cos(\omega t - kz - \varphi)\vec{a}_y \tag{4.2}$$

where

$$\begin{cases} E_x(z,t) = E_{x0}\cos(\omega t - kz) \\ E_y(z,t) = E_{y0}\cos(\omega t - kz - \varphi) \end{cases} \tag{4.3}$$

The $E_x(z,t)$ and $E_y(z,t)$ can be concluded by the relation Eq. (4.3)

$$\left[\frac{E_x(z,t)}{E_{x0}}\right]^2 + \left[\frac{E_y(z,t)}{E_{y0}}\right]^2 - \frac{2E_x(z,t)E_y(z,t)}{E_{x0}E_{y0}}\cos\varphi = \sin^2\varphi \tag{4.4}$$

The distance of the two slots is $\lambda_g/4$ along the feed direction, which is equivalent to the phase difference $\varphi = \pm 90°$. Assuming the amplitude of E-fields are $E_{x0} = E_{y0} = E_0$, then

$$E_x^2(z,t) + E_y^2(z,t) = E_0^2 \tag{4.5}$$

Equation (4.5) shows that the amplitude of $\vec{E}(z,t)$ is a constant, and it doesn't vary with time, and the angles between +X axis direction and the polarization direction of $\vec{E}(z,t)$ is

$$\alpha = \arctan\frac{E_0\cos(\omega t - kz \pm \frac{\pi}{2})}{E_0\cos(\omega t - kz)} = \arctan[\nu\tan(\omega t - kz)] = \nu(\omega t - kz) \tag{4.6}$$

For any given position, z is a constant, the direction of $\vec{E}(z,t)$ rotates with the constant angular frequency ω with time, this is the circular polarization wave. In Fig. 4.2, when the field is excited as fed direction 1, the phase of E_y E_x is ahead 90° than the phase of E_y, which means $\varphi = 90°$ and $\alpha = \omega t$. The electric field vector rotates in anti-clockwise direction, and it is RHCP wave. On the contrary, when the fields are excited as fed direction 2, the phase of E_x lagged 90° than the phase of E_y, so $\varphi = -90°$ and $\alpha = -\omega t$. The electric fields vector rotate in the clockwise direction, and it is LHCP wave. Comparing Fig. 4.2 with the slots array arrangement in Fig. 4.1, the feed direction 1 and feed direction 2 are corresponding to the RHCP feed port and LHCP feed port respectively.

Since the distances between the adjacent slots are different, the amplitudes of the electric fields from two slots are not equal ($E_{x0} \neq E_{y0}$). So, the distance difference will affect the axial ratio of the main beam.

4.3 Antenna Performance Analysis

The experiment results and simulation results of the designed omnidirectional dual-CP antenna (ODCPA) are compared and discussed in this section. For the dual-CP performances excited from different ports, the LHCP character and RHCP character of the ODCPA are measured one by one. Firstly, the LHCP port (port 1) is fed and one 50 Ω load is connected to the RHCP port (port 2), so that the antenna radiates

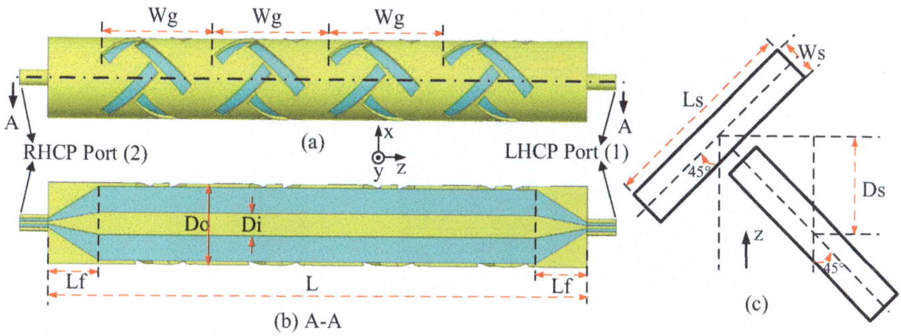

Fig. 4.1 Geometry of the omnidirectional dual circularly polarized (CP) antenna. **a** Side view. **b** A-A section view. **c** Geometry of the slot pairs [13]. © 2015 IJAP

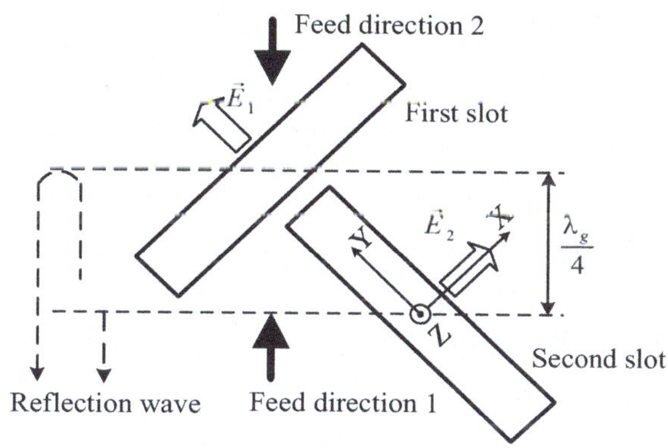

Fig. 4.2 The adjacent perpendicular slot pairs [13]. © 2015 IJAP

the LHCP wave and achieve LHCP farfield in the experiment. Secondly, exchange the status of these two ports, port 1 connects to one 50 Ω load and port 2 is excited, then the RHCP field is radiated, and realize the RHCP far field to measure. The manufactured ODCPA and experiment scene in chamber is shown in Fig. 4.3.

Figure 4.4 shows the measured *S*-parameters curves. The practical available work band of the manufactured ODCPA with $|S_{11}| < -10$ dB and $|S_{22}| < -10$ dB is about 16.4% in 5.05–5.95 GHz, and the isolation between the LHCP port and RHCP port is higher than 15 dB in the same work band. It is clear that the return losses of the two ports in the ODCPA are very small, and the isolation between port 1 and port 2 is high enough, so, the designed ODCPA performs well from these two CP polarization ports.

The normalized far field patterns of the co-polarization and cross-polarization in azimuth plane (*xoy*-plane) and elevation plane (*xoz*-plane) from port 1 and port 2 are presented in Fig. 4.5 respectively at 5.1 GHz, 5.5 GHz and 5.9 GHz. The measured far fields of LHCP or RHCP in azimuth plane (*xoy*-plane) are circles with little burrs, which means that the designed ODCPA has good omnidirectional radiation character in azimuth plane. The measured radiation patterns of LHCP or RHCP in elevation

Fig. 4.3 Photograph of the omnidirectional CP antenna being tested in the anechoic chamber [13]. © 2015 IJAP

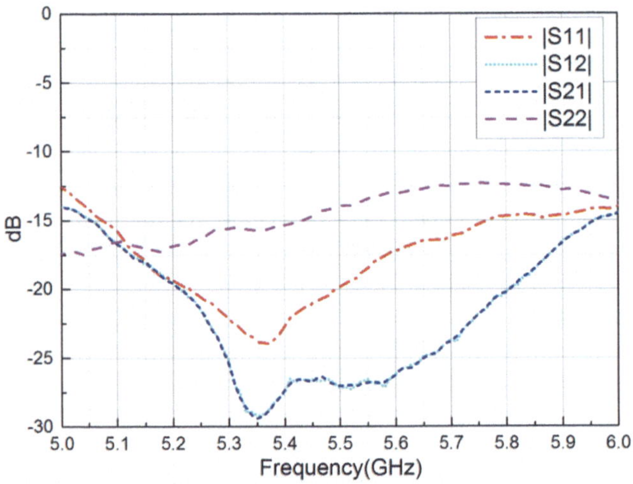

Fig. 4.4 Measured *S*-parameters results [13]. © 2015 IJAP

plane (*xoz*-plane) have clear main beams directing to $\theta = 90°$, which means that the ODCPA really radiates in azimuth plane (*xoy*-plane) with omnidirectional beam. Secondly, the measured far fields are almost consistent with the simulated results besides a little discrepancy. Similar as the results error of the OSA in Chapter 3, the reasons for the error are mainly the impure medium with permittivity error and manufacturing error from the complicated structure.

Thirdly, the cross-polarization beams are lower 15 dB than the co-polarization main beams in azimuth plane (*xoy*-plane) or elevation plane (*xoz*-plane) from port 1 and port 2 respectively.

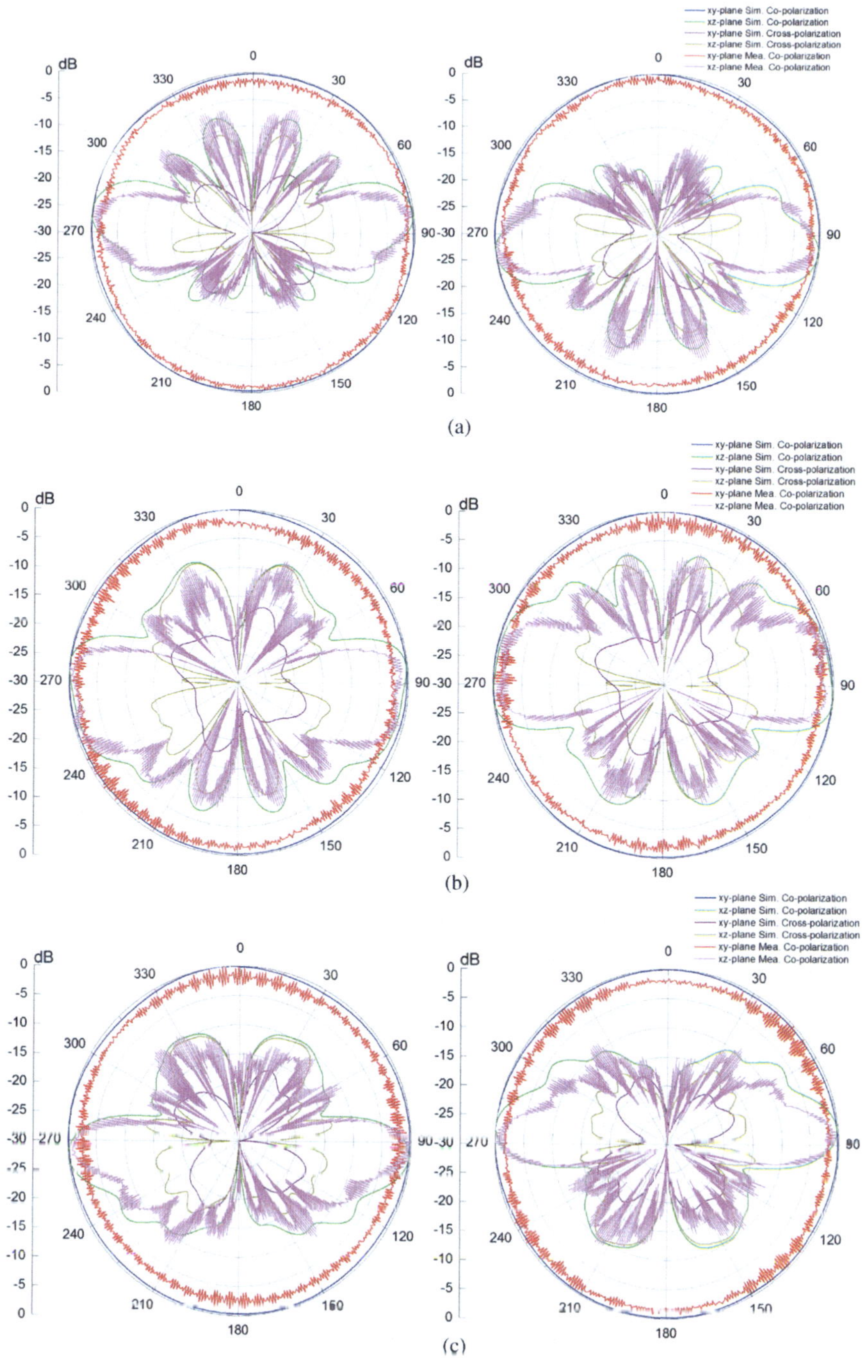

Fig. 4.5 Simulated and measured LHCP (in left) and RHCP (in right) normalized co-polarization and cross-polarization radiation patterns in the xy-plane (omnidirectional plane) and xz-plane, **a** 5.1 GHz, **b** 5.5 GHz, **c** 5.9 GHz [13]. © 2015 IJAP

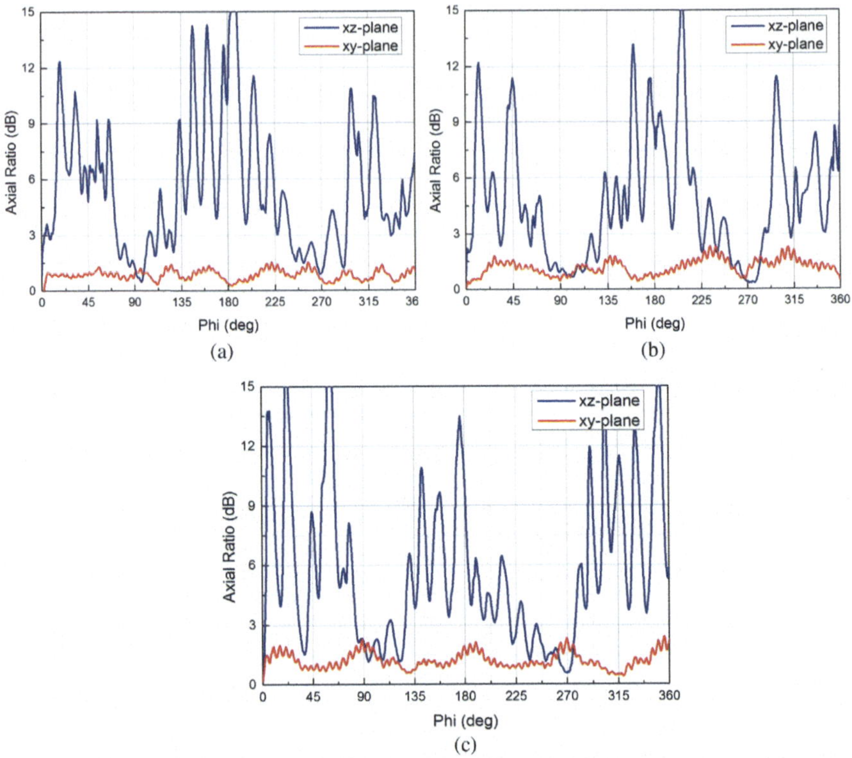

Fig. 4.6 Measured axial ratio patterns in xy-plane and xz-plane, **a** 5.1 GHz, **b** 5.5 GHz, **c** 5.9 GHz [13]. © 2015 IJAP

The measured axial ratio curves in azimuth plane (*xoy*-plane) or elevation plane (*xoz*-plane) of LHCP cases at 5.1 GHz, 5.5 GHz and 5.9 GHz are given in Fig. 4.6. It is clear that the axial ratio curves of the main beam in azimuth plane (*xoy*-plane) are less than 3 dB at these 3 frequency points. In another side, comparing the radiation pattern elevation plane (*xoz*-plane) and the axial ratio curves of LHCP, the axial ratio is lower than 3 dB during the HPBW (half power beam width). It again confirms that the designed ODCPA works well in circular polarization. It is similar to the RHCP cases excited form port 2.

In Fig. 4.6, there are small ripples in the axial ratio curves in azimuth plane (*xoy*-plane). As discussed, the principle of ODCPA in Sect. 4.2, the circular polarization performance in the direction of $\theta = 90°$ is very good with minimum axial ratio value, but the CP performance will deteriorate in non-vertical directions. For one round of radiation slots unit, there are four slots pairs in the outer conductor shell around the coaxial cylinder, but they are not coplanar. These periodical structure variances around the coaxial cylinder will generate small irregular wave deformation, and it appears periodical small ripples in the far fields and axial ratio curves in azimuth plane.

Figure 4.7 shows the measured gain curves and average axial ratio curves in the azimuth plane (*xoy*-plane). The measured gain is larger than 4 dBi during 5.0–6.0 GHz, and the average axial ratio (AAR) value is lower than 1.5 dB in the azimuth plane (*xoy*-plane) in the same bandwidth.

The simulated total efficiency of the designed ODCPA is given in Fig. 4.8. It can be seen that the efficiency of ODCPA is in 86.9–93.5% during 5.1–5.9 GHz, which means that the work efficiency of the designed ODCPA is high in the work band.

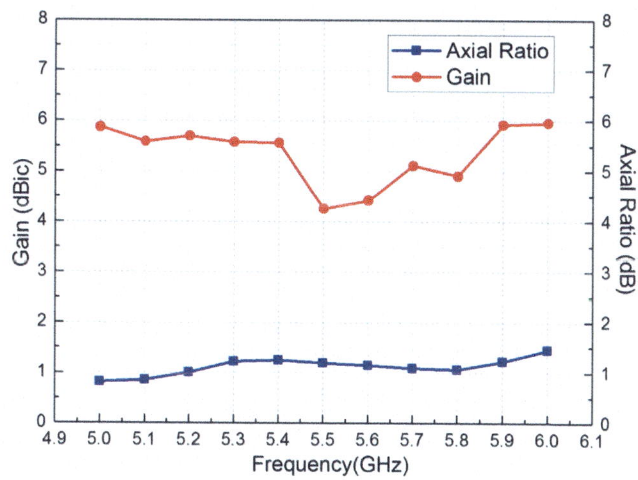

Fig. 4.7 Measured gain and average axial ratio in the omnidirectional plane [13]. © 2015 IJAP

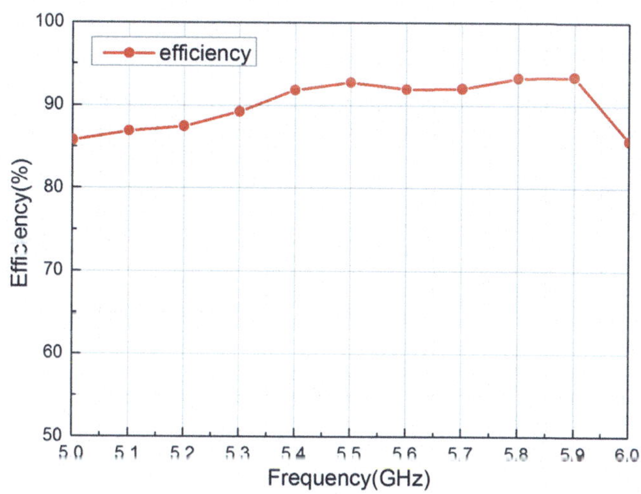

Fig. 4.8 The total efficiency results of the proposed antenna [13]. © 2015 IJAP

4.4 Summary

The omnidirectional slots array antenna with dual-CP character is presented. The antenna works from the slots array in the outer conductor shell around the coaxial cylinder. It has two ports in the ends of the coaxial cylinder to excite LHCP and RHCP respectively. If the down port is excited and the top port connects to one 50 Ω load to match, the slot array radiates the LHCP wave. Conversely, the top port is excited and the down port connects to a 50 Ω load, the RHCP wave is radiated too. The available impedance bandwidth of the ODCPA with $|S_{11}| < -10$ dB and $|S_{22}| < -10$ dB is about 16.4% in 5.05–5.95 GHz, and the isolation between the LHCP port and RHCP port is higher than 15 dB in the same work band. The designed ODCPA can wok in LHCP mode or RHCP mode with omnidirectional radiation pattern in azimuth plane. and the average axial ratio (AAR) is below 1.5 dB in azimuth plane. The gain in LHCP or RHCP mode is larger than 4 dBi with radiation efficiency over 86.9% during the available work band.

References

1. W. Croswell, C. Cockrell, An omnidirectional microwave antenna for use on spacecraft. IEEE Trans. Antennas Propag. **17**(4), 459–466 (1969)
2. K. Sakaguchi, N. Hasebe, A circularly polarized omnidirectional antenna, in *Proceedings of the 8th International Conference on Antennas and Propagation (ICAP'93)* (1993), pp. 477–480
3. C.Y. Yu, T.H. Xu, C.J. Liu, Design of a novel UWB omnidirectional antenna using particle swarm optimization. Int. J. Antennas Propag. **2015**, Article ID 303195, 7 pages (2015)
4. A. Narbudowicz, X.L. Bao, M.J. Ammann, Miniaturization of omnidirectional circularly polarized antennas, in *Proceedings of the 8th European Conference on Antennas and Propagation (EuCAP'14)*, April 2014, pp. 2919–2923
5. M. Khalily, M.R. Kamarudin, M. Mokayef, M.H. Jamaluddin, Omnidirectional circularly polarized dielectric resonator antenna for 5.2 GHz WLAN applications. IEEE Antennas Wirel. Propag. Lett. **13**, 443–446 (2014)
6. X.D. Bai, X.L. Liang, M.Z. Li, B. Zhou, J. Geng, R. Jin, Dual-circularly polarized conical-beam microstrip antenna. IEEE Antennas Wirel. Propag. Lett. **14**, 482–485 (2015)
7. D. Yu, S.-X. Gong, Y.-T. Wan, W.-F. Chen, Omnidirectional dual-band dual circularly polarized microstrip antenna Using TM01 and TM02 modes. IEEE Antennas Wirel. Propag. Lett. **13**, 1104–1107 (2014)
8. X.-L. Quan, R.-L. Li, Broadband dual-polarized omnidirectional antennas, in *Proceedings of the IEEE Antennas and Propagation Society International Symposium (APSURSI'12)*, IEEE, Chicago, Ill, USA, July 2012, pp. 1–2
9. B. Zhou, J.P. Geng, X.D. Bai, L. Duan, X. Liang, R. Jin, An omnidirectional circularly polarized slot array antenna with high gain in a wide bandwidth. IEEE Antennas Wirel. Propag. Lett. **14**, 666–669 (2015)
10. K. Iigusa, T. Teshirogi, M. Fujita, S.-I. Yamamoto, T. Ikegami, A slot-array antenna on a coaxial cylinder with a circularly polarized conical beam. Electron. Commun. Jpn. Part I: Commun. **83**(3), 74–87 (2000)
11. M.E. Bialkowski, P.W. Davis, slot-array antenna with a broadened beam. Microw. Opt. Technol. Lett. **27**(2), 98–101 (2000)

12. B. Zhou, High gain omnidirectional circularly polarized antenna with slots on coaxial waveguide. Mater thesis, Shanghai Jiao Tong University, 2016
13. B. Zhou, J. Geng, Z. Li, W. Wang, X. Liang, R. Jin, Dual circularly polarized omnidirectional antenna with slot array on coaxial cylinder. Int. J. Antennas Propag. **2015**, Article ID 127820, 7 pages (2015)

Chapter 5
Dual CP Polarization Diversity and Space Diversity Antennas Enabled by a Compact T-Shaped Feed Structure

5.1 Introduction

With the rapid growth of the demand for wireless transmission rate and capacity, wireless communication systems are developed quickly, and the antenna is not only to transmit or receive signal, but also increase the system performances. Such as polarization diversity (PDA) and spatial diversity (SDA) antennas, has been strong powerful to achieve large cover and high-speed transmission. Polarization diversity is very useful to anti channel fading in the complex and vicious electromagnetic environments. Polarization diversity can support different polarization states to transfer independent EM signal in the same wireless channel, which increases the channel efficiency and capacity [1–3]. In another side, spatial diversity or pattern diversity makes the space coverage more flexible. Spatial diversity means more space channels can be used to transfer EM signal, and finally combined to achieve high quality signal or simultaneously to transfer in different scenarios. Circular polarization (CP) is very good to receive the same circular polarization, elliptical polarization or linear polarization EM wave, which is well to realize polarization diversity. Spatial diversity antenna can use different beams to point to different directions, which balances the wider coverage, high gain and suppressing the interference between different ports and modes.

To the multi-function mobile terminals, the PDA and SDA antennas are very useful, they do not only connect to neighbouring nodes, but also communicate with unmanned aerial vehicles (UAVs) or satellite [4, 5]. These antennas usually are switchable antenna with one port or have two ports to excite two orthogonal mode in the antenna. To the mobile wireless communication platform, it is necessary to have the power of hemispherical EM coverage, large angle CP beams with high gain e.g., mobile communication in ad hoc network [4–6]. In fact, there are many research being reported on diversity to overcome multipath fading and increase the channel capacity [7–9].

In [10], an antenna combining the monopole and loop radiators to radiate omnidirectional RHCP wave in the azimuthal plane was proposed. The bandwidth with AR

J. Geng, *Omnidirectional Slots Antenna*,
https://doi.org/10.1007/978-981-15-9089-4_5

< 3 dB is 17.5%, over 1.46–1.74 GHz. Its gain is 0.86 dBic. In [11], a broadband dual-CP omnidirectional antenna was designed by combining LHCP and RHCP elements on the downside and upside of a cylinder [11]. A wide band compact patch antenna with low-profile was presented in [12]. It realized reconfigurable polarization performance with PIN diodes on slot. Certainly, the insertion loss was increased, and the additional complex biasing circuits was introduced too.

Omnidirectional CP antenna with high gain in the azimuth plane is very useful to vehicle communication on the ground. We have designed a series of slot array antenna in the outer conductor shell of the coaxial cylinder in [13–18]. Omnidircctional slots antenna with CP character was firstly presented in [13] based on a coaxial cylinder waveguide with slot arrays around its axis. There are four perpendicular slots pairs divided by a half medium wavelength interval on the outer conductor shell to radiate omnidirectional CP wave. The available band with $|S11| < -10$ dB and AR < 3 dB is 5.1–5.9 GHz, and its peak gain is 6.5 dBic. In [14], higher gain is achieved by increasing slots. The omnidirectional slot antenna with CP characters proposed in [13, 14] can only work in RHCP mode with an individual feeding port.

In further, an omnidirectional slot array antenna with dual-CP character was designed in [15]. It has two ports on two side of the coaxial cylinder with slots array on the outer conductor shell. One port excite omnidirectional LHCP wave, and another port to excite omnidirectional RHCP wave, respectively. However, the LHCP port is on the top of the coaxial cylinder, and the feed wire to the top port from the outside of the antenna deteriorates the omnidirectional dual-CP radiation patterns when it was vertically mounted on the vehicle metallic top, so that it cannot achieve realize omnidirectional radiation and its CP characteristics are broken in some directions.

The omnidirectional slots array antenna was extended in [17] to a related truncated circular cone with slot array. It radiates the conical beam with CP character by feeding from a bottom port, the guided wave running its throat ring, and then radiating out from the slots array. The dip angle of the conical beam was decided by the vertex angle of the outer conductor and the distribution of the slot arrays. In [18], we proposed a T-shaped feed structure to excite the multi-port antenna system. There are two independent channel with high isolation and very small insertion loss in both channel.

In further, an improved coaxial T-shaped feed structure (TFS) with 4 ports is designed in detail in this chapter. It can feed the omnidirectional CP slot array antenna (OCPSA) to realize polarization diversity antenna and spatial diversity antenna. The TFS is compact and can work with two independent channels with enough isolation.

When the TFS connects with the omnidirectional CP slot array antenna (OCPSA)to realize omnidirectional radiation with dual-CP character, it is called improved DCPOA or polarization diversity antenna (PDA), which is fed by the independent channels of the TFS, and radiate LHCP wave or RHCP wave, respectively. The improved DCPOA is very convenient realize CP diversity in the azimuth plane when the antenna is perpendicularly on the roof of the vehicle.

When the TFS combined with a single port feeding omnidirectional slot array CP antenna in [13–16], and another channel connects with a low-profile directional antenna on the top of the cylinder by an internal thin coaxial cable (ITCC) embedded in the center of the structure, it can realize a spatial diversity antenna (SDA) in wide band. One channel of the TFS feed the omnidirectional slot array CP antenna to achieve omnidirectional radiation in azimuth plane, and another channel excites the upwards radiation field. The proposed PDA and SDA all were manufactured, and the experiment results were close to the simulation results.

5.2 T-Shaped Feed Structure

Here, the commercial software CST Microwave Studio is applied to design and optimize the improved dual-CP omnidirectional antenna, which is based on the time domain finite integration technique (FIT) method and finite difference time domain (FDTD) algorithms. The finite element (FEM) method was used to computation in frequency domain simulations in CST too [19, 20].

Compare to the reported microwave networks with multiport, the proposed TFS is compact with two built-in channels with high isolation. The TFS can be applied to conveniently feed PDA or SDA from the down side, so that remove the defects that the outside feeding cable to top port deteriorates the pattern roundness in azimuth plane, and the PDA or SDA is easily installed on the roof of the vehicles.

5.2.1 TFS Geometry Structure

The TFS structure is shown in Fig. 5.1, which developed from the coaxial cable and transmission line with branches [29]. There are two isolated physical channels to transfer the independent electromagnetic waves. One is a thin built-in channel of the internal thin coaxial cable (ITCC) with t port 1 and port 3, which is uniform with a 50 Ω characteristic impedance. This channel structure is given in Fig. 5.1b. A standard SMA adaptor links with Port 1. Port 3 directly connects the antenna. Another channel links port 2 and port 4, as shown in Fig. 5.1c. It in fact is a bent coaxial transmission line with a short branch to adjust the impedance match. The entrance coaxial wire to port 2 is a tapered coaxial line with a fixed radius ratio and characteristic impedance. Port 2 connects a standard SMA adaptor. Port 4 connects the horn-shaped coaxial line part. In another side, port 4 combines with the coaxial cylinder with slots array on the shell. The microwave channel between port 2 and port 4 can be equivalent to microwave circuit, depicted in Fig. 5.1d. The bottom branch is the shortened branch, and the end ring (marked by the red frame in Fig. 5.1a, b) reflects the EM waves to change the incident wave phase when the l_1 is adjusted to ensure that the combined electromagnetic signal has the same phase at port 4. In fact, the vertical distance l_1

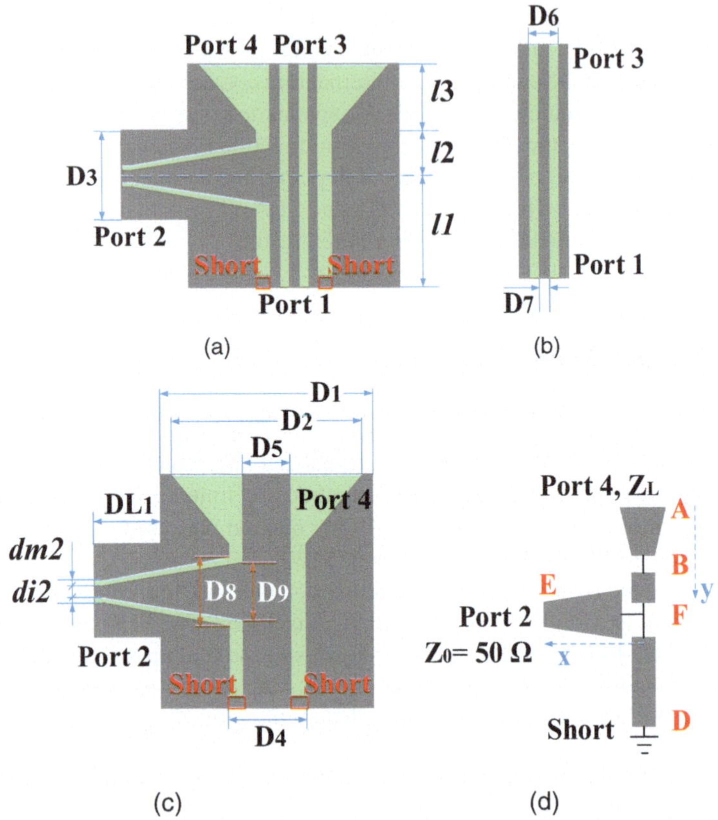

Fig. 5.1 The T-shaped feed structure (TFS). **a** A cross-sectional view through a vertical cut. **b** Port 1 is connected directly to port 3 by a coaxial line. **c** The channel connects port 2 and port 4 with a bent coaxial cable and a shortened parallel horn-shaped coaxial branch. **d** Equivalent circuit of (**c**). The signal enters from port 2 and flows to port 4, when port 1 is shortened. Figure reproduced with permission from: [29], © 2019 IEEE

is the key to rise the transferring efficiency in the channel 2 from port 2 to port 4, and can better the EM field distribution around port 4.

The inner space of the TFS is filled by Teflon with $\varepsilon_r = 2.1$.

5.2.2 TFS Impedance Transformation

In fact, the key of the TFS is the impedance and efficiency of the channel 2 between port 2 and port 4. Especially the shortened branch l_1 is essential to the performance of the system, and port 4 combines the large aperture coaxial cylinder with variable size. So, the organic combination with optimized size of the components is important to get low insertion loss and good impedance matching.

We suppose Z_L as the load to port 4 in Fig. 5.1d. The characteristic impedance of the coaxial cable AB with tapered structure can be expressed as

$$Z_c(y) = \frac{60}{\sqrt{\varepsilon_r}} \ln \frac{2}{D_5}[\frac{D_2 - D_4}{2l_3}(l_3 - y) + D_4] \tag{5.1}$$

Here, Teflon is filled the space of the coaxial cable, and ε_r is the permittivity. All the size parameters are marked in Fig. 5.1. The input impedance at $y + dy$ in part AB can be written as Eq. (5.2) by the principle of the impedance transmission with tapered structure.

$$Z_{in}(y + dy) = Z_c(y)\frac{Z_{in}(y) + jZ_c(y)\tan(\beta dy)}{Z_c(y) + jZ_{in}(y)\tan(\beta dy)} \tag{5.2}$$

In further, when $dy \to 0$, Eq. (5.2) can be approximated as

$$Z_{in}(y + dy) \cong Z_{in}(y) + Z_{in}^{'}(y)dy \tag{5.3}$$

From A to B, the terminal load impedance $Z_L = Z_{in}(y = 0)$ can be transformed to be

$$Z_{in}(y = l_3) = Z_{in}(y = 0) + \int_0^{l_3} Z_{in}^{'}(y)dy \tag{5.4}$$

For part BF, the uniform coaxial line, the characteristic impedance can be directly achieved

$$Z_c(BF) = \frac{60}{\sqrt{\varepsilon_r}} \ln \frac{D_2}{D_5} \tag{5.5}$$

And the impedance $Z_{in}(y = l_3)$ at B, the is transformed to F

$$Z_{F1}(y = l_3 + l_2) = Z_c(BF)\frac{Z_{in}(y = l_3) + jZ_c(BF)\tan(\beta l_2)}{Z_c(BF) + jZ_{in}(y = l_3)\tan(\beta l_2)} \tag{5.6}$$

Part DF is shorted at D, which is the uniform coaxial transmission line with the characteristic impedance $Z_c(DF)$ as $Z_c(BF)$. So, the input impedance at position F from this shortened branch is transformed by part DF to be Z_{F2}.

$$Z_{F2} = jZ_c(DF)\tan(\beta l_1) \tag{5.7}$$

Here, Z_{F1} and Z_{F2} are a parallel connection, and $Z_F = Z_{F2}//Z_{F2}$.

Part FE is tapered coaxial transmission wire too. Its characteristic impedance can be written as

$$Z_c(x) = \frac{60}{\sqrt{\varepsilon_r}} \ln \frac{D_8 \cdot (D_1 - D_5 + 2DL_1)}{D_9 \cdot (D_1 - D_4 + 2DL_1)} = Z_{c0} \tag{5.8}$$

Similar as the transmission case of Eq. (5.2) of the impedance transformation function to the tapered transmission line, the impedance at $x + dx$ in part FE can be written as

$$\begin{aligned} Z_{in}(x + dx) &= Z_{c0} \frac{Z_{in}(x) + jZ_{c0} \tan(\beta dx)}{Z_{c0} + jZ_{in}(x) \tan(\beta dx)} \\ &= Z_{in}(x) + Z'_{in}(x)dx \end{aligned} \tag{5.9}$$

The impedance Z_F at position F is transformed to point E,

$$Z_{in}\left(E, x = DL_1 + \frac{D_1}{2}\right) = Z_F + \int_0^{DL_1 + \frac{D_1}{2}} Z'_{in}(x)dx \tag{5.10}$$

Then 50 Ω SMA adaptor at port 2 is connected to position E,

$$Z_{in}\left(E, x = DL_1 + \frac{D_1}{2}\right) = 50\,\Omega \tag{5.11}$$

For a given load Z_L at port 4, the size parameters values of TFS can be determined by solving the above equations. Here, suppose $Z_L = 100\ \Omega$ and the initial parameter size of the TFS can be calculated. A count part EM model of TFS is established, simulated and optimized in CST.

5.2.3 TFS Performance by Simulation

The simulation results of TFS are shown in Fig. 5.2. It can be seen that the shortened branch length l_1 affects the S parameters in the entire operational band. As discussed above, the transformation impedance is adjusted by the shortened branch length l_1. From the results in Fig. 5.2, the optimized shortened length is $l_1 = 17$ mm with the simulated return loss at port 2, $|S_{22}| < -10$ dB over 2.31–8.95 GHz. The transmission loss $|S_{42}|$ is less than 0.5 dB, and the isolation coefficient $|S_{12}|$ between the two input port 1 and port 2 is below −30 dB between 2.33 and 8.93 GHz.

When l_1 is 12 mm, $|S_{22}|$ will deteriorate it is clear that if l_1 deviates from 17 mm, $|S_{22}|$ and $|S_{42}|$ will deteriorate seriously. The length of the coaxial horn line affect the performance too, which controls the phase difference between these two EM channels. It can be seen that the available impedance band of the TFS will deviate the optimized band and become narrow if l_1 leave from the optimized value, 17 mm.

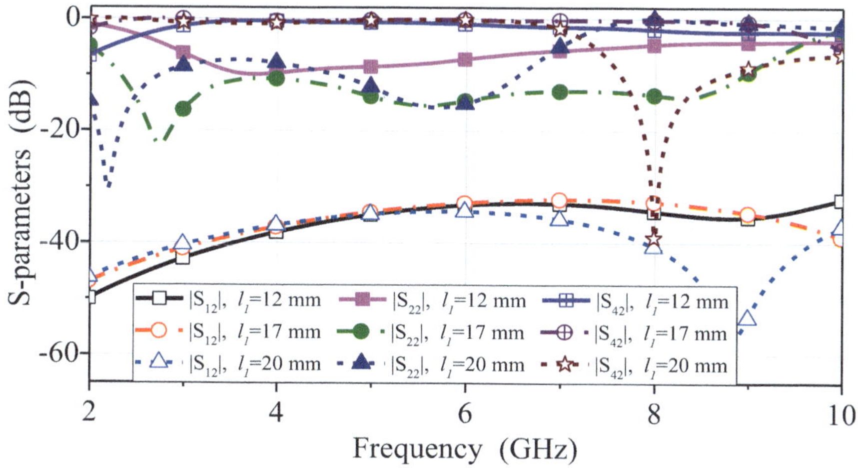

Fig. 5.2 S parameters when l_1 = 12, 17, and 20 mm. Figure reproduced with permission from: [29], © 2019 IEEE

*DL*1, the length of coaxial tapered line, is another key parameter too. As shown in Fig. 5.3, *DL*1 affects the impedance transformation result, so that affects the transferring performance of channel between the port 2 and port 4. When *DL*1 leaves from 7 mm, become smaller or larger, the available impedance band of the channel between port 2 and port 4 in the TFS will become worse, and the operation bandwidth will become narrower from the original band.

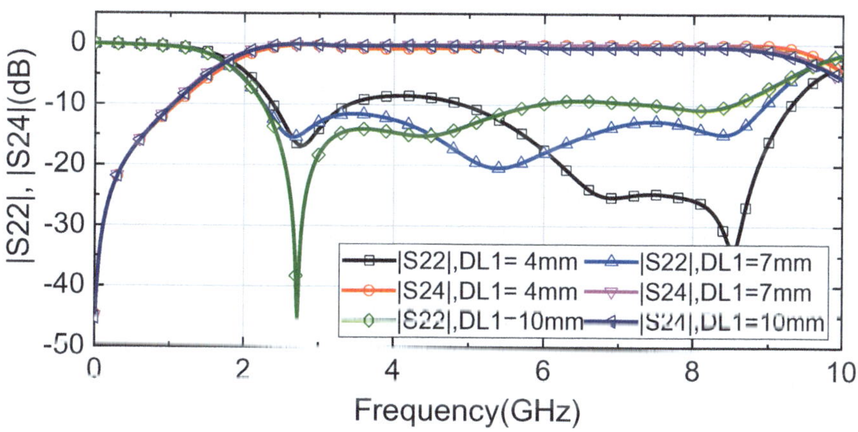

Fig. 5.3 S parameter when *DL*1 = 4, 7, 10 mm

5.3 The Improved Dual CP Omnidirectional Antenna Fed by TFS

Polarization diversity antenna (PDA) means the antenna can work in different polarization states, and it can choose its polarization if it is necessary. PDA is very powerful to overcome de-polarization, improve the stability of the channel, and raise the capacity of the communications.

In another side, the omnidirectional CP slots array antenna (OCPSA) proposed in Chap. 4 can support dual-CP performance, and fed from bottom port and top port by outside cable [15]. However, the outside feed cable affects the roundness of the omnidirectional pattern and deteriorates the CP performance in some direction. Here, the TFS is a good solver to feed the OCPSA without these defects, and provides an improved feeding of the DCPOA.

5.3.1 The Principle and Structure of the Improved DCPOA

The coaxial cylinder structure with slot arrays and the TFS are the main parts of the improved DCPOA. The omnidirectional LHCP or RHCP radiation performances are realized by exciting different ports [15]. The improved DCPOA with TFS feeding is shown in Fig. 5.4. The feed network principle of TFS is illustrated in Fig. 5.4a. Feed from port 1 to port 3, and then excites the LHCP antenna; Feed from port 2 to port 4, and excites the RHCP antenna. In Fig. 5.4b, port 1 on the bottom is directly fed to excite slots array to radiate the LHCP wave, and port 2 on the lower side is fed to transfer the EM wave into the bend coaxial transmission line to excite the slots array to radiate the RHCP wave. In this construct, the ITCC take the place of the core metal wire of the OCPSA in [13], the inner core line of the ITCC is straightly extended to connect the top cap of the coaxial cylinder in Fig. 5.4c. Between the top cap and the outer conductor of the ITCC, there is a small gap ring 's', which is the throat to the EM signal. When the EM field is excited at port 1, it run into the ITCC to the top of the antenna with TEM, then turns back and enters into the small gap ring 's', down flow into the external cavity of the coaxial slots array to radiate the LHCP wave, as shown in Fig. 5.4c. Similarly, the EM field is excited from port 2, then runs through port 4, and propagates into the original coaxial slot array antenna with TEM wave, and transfers from the bottom to the top with radiating RHCP fields. This kind of nested coaxial structure includes the slot array around the coaxial cylinder to radiate and TFS to feed from down side, so that it is compact with high performances.

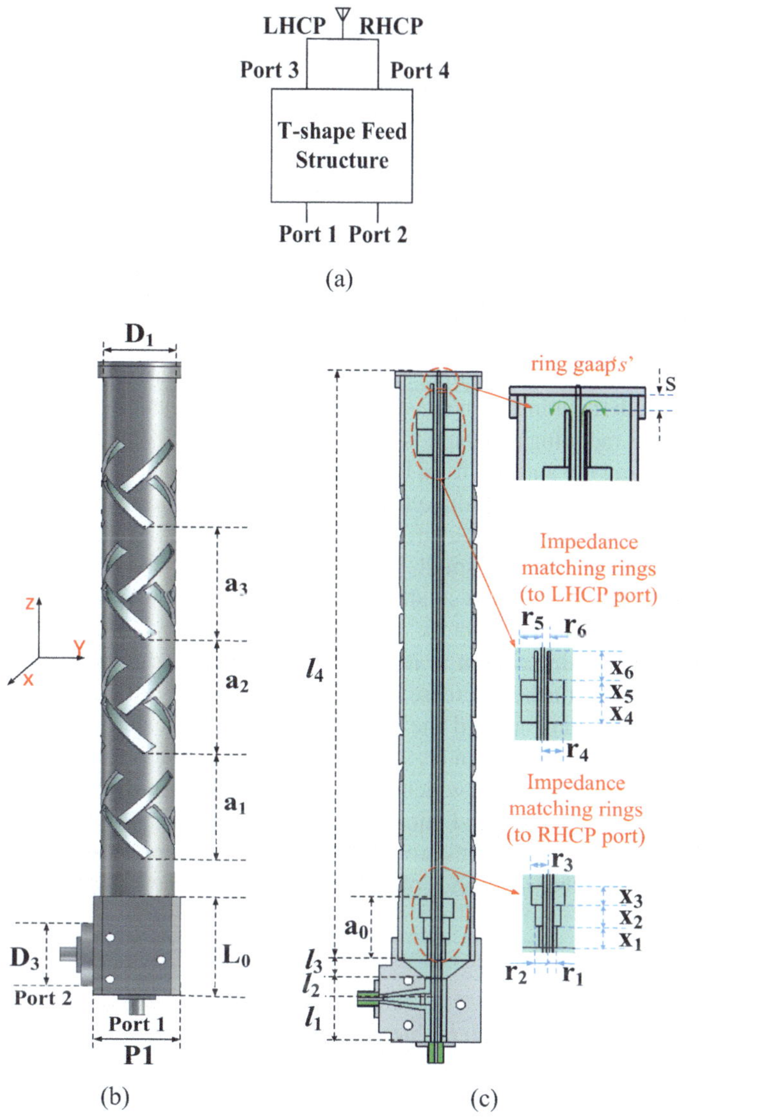

Fig. 5.4 The T-shaped fed dual-CP slot arrays antenna. **a** Overview diagram. **b** 3D view. **c** Cross-sectional view through a vertical cut. Figure reproduced with permission from: [18, 29], © 2019 IEEE

5.3.2 Performance of the Improved DCPOA

The port 1 on the bottom is LHCP port and the port 2 on the side is RHCP port. Both of the two ports character impedance are designed as 50 Ω, so that they can connect SMA adaptor directly. When LHCP port is excited and RHCP port terminated with a 50 Ω load, the antenna generates the LHCP radiation, whereas the RHCP radiation is excited when RHCP port is excited and LHCP port terminated with a 50 Ω load.

One coaxial line (inner coaxial line) run through the antenna. It serves as a channel for EM signal to go to the top of the antenna. The outer conductor of the coaxial line and the outer conductor of the antenna also serve as another coaxial line (outer coaxial line) for EM signal to pass through from top of the antenna to the bottom. The detail of the connection between the two coaxial lines can be seen in Fig. 5.4c. The inner conductor of the ITCC is extracted $s = 2.9$ mm to connect the top metal cap of the antenna. As is shown in Fig. 5.4a, when excite the antenna form LHCP port, the energy go through the inner coaxial line from bottom to the top until it arrive the metal cap and runs out from the ring gap between the top cap and the out conductor of the inner coaxial line, and then turns around passing through the outer coaxial line with radiating from the slots pairs to the air. It can be observed that most of the energy is radiating and there is few amount of energy going to the RHCP port.

When exciting the antenna from RHCP port, the EM signal just goes through the outer coaxial line as well as radiate from the slots pairs to the air, which is shown in Fig. 5.5b, it shows that the three circle of slots pairs at front radiate a lot energy but the last one radiate relatively small, which means the last make a little contribution to the gain of antenna and even small energy is leak to the other port, so we got good isolation result.

What the above description is just the feeding way to the LHCP antenna and RHCP antenna. To further increase the radiation efficiency of LHCP antenna, we insert three metal rings around the inner coaxial line to match the impedance as shown in Fig. 5.4c, similar as in [1, 2], they are close to the top end, and their length are $x4$, $x5$ and $x6$ respectively, which are calculated and optimized by the equivalent circuit of the LHCP antenna radiation impedance. Similar as the RHCP antenna, we insert the other three metal rings around the inner coaxial line close to the T-shaped connector to match the impedance of RHCP antenna, and their length are $x1$, $x2$ and $x3$ respectively.

5.3.3 DCPOA Parametric Study and Analysis

There are some key parameters that have great influence on the antenna performance. The length of the slots mainly determines the radiation frequency of the DCPOA. The ideal length of the slots is half wavelength same as in [13]. In order to have a wider bandwidth and considering the power distribution among the four circles of the slot pairs, the intervals between these four rounds slots are optimized with setting $a1 < a2 > a3$. So rather excite from top or bottom of the out conductor, the slot pairs are almost the same, which is good for balancing the performance the RHCP radiation and LHCP radiation. On the other hand, it's good for equal radiation strength between the four rounds slot pairs and get good directivity in azimuth plane [15].

The isolation between the RHCP port and LHCP port is essential to the functionality of polarization diversity. We should make the isolation bandwidth coincide

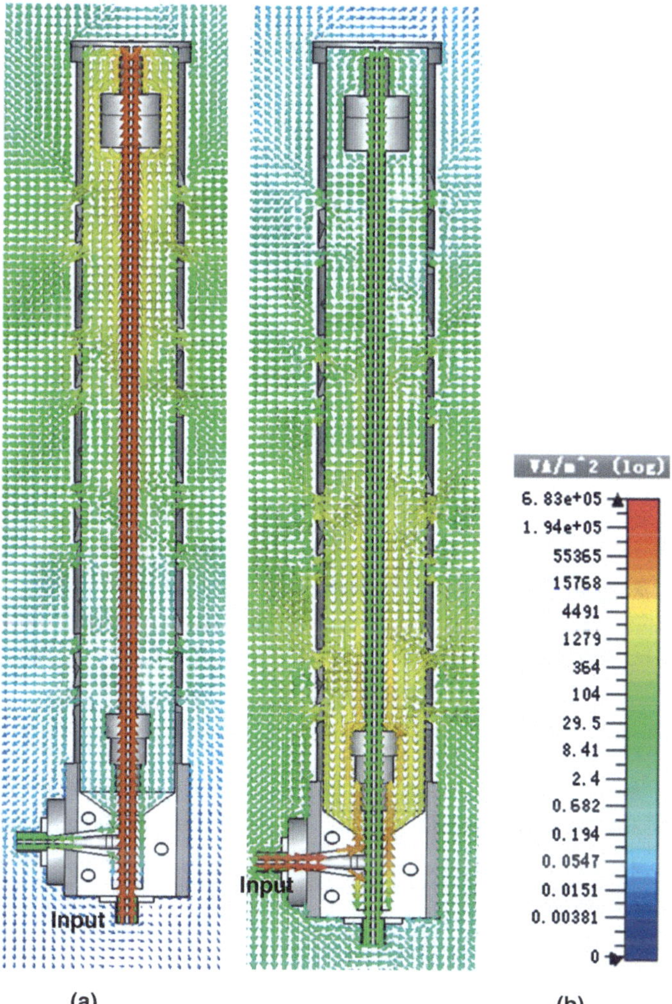

Fig. 5.5 Power flow in the structure. **a** When fed from the LHCP port 1. **b** When fed from the RHCP port 2. Figure reproduced with permission from: [29], © 2019 IEEE

exactly with the impedance bandwidth. The key parameter to adjust the isolation bandwidth is the width of the slots. By reasonably adjusting the width of the slots, good isolation property is achieved.

(i) Impedance matching rings in the improved DCPOA

The impedance matching metal rings is the key to match wide bandwidth [13]. The RHCP impedance match rings are shown in Fig. 5.4c, which can be achieved by optimizing the equivalent microwave circuits with parallel branches, which is very important to the impedance bandwidth of port 2. In Fig. 5.6a, the |S22| will be serous

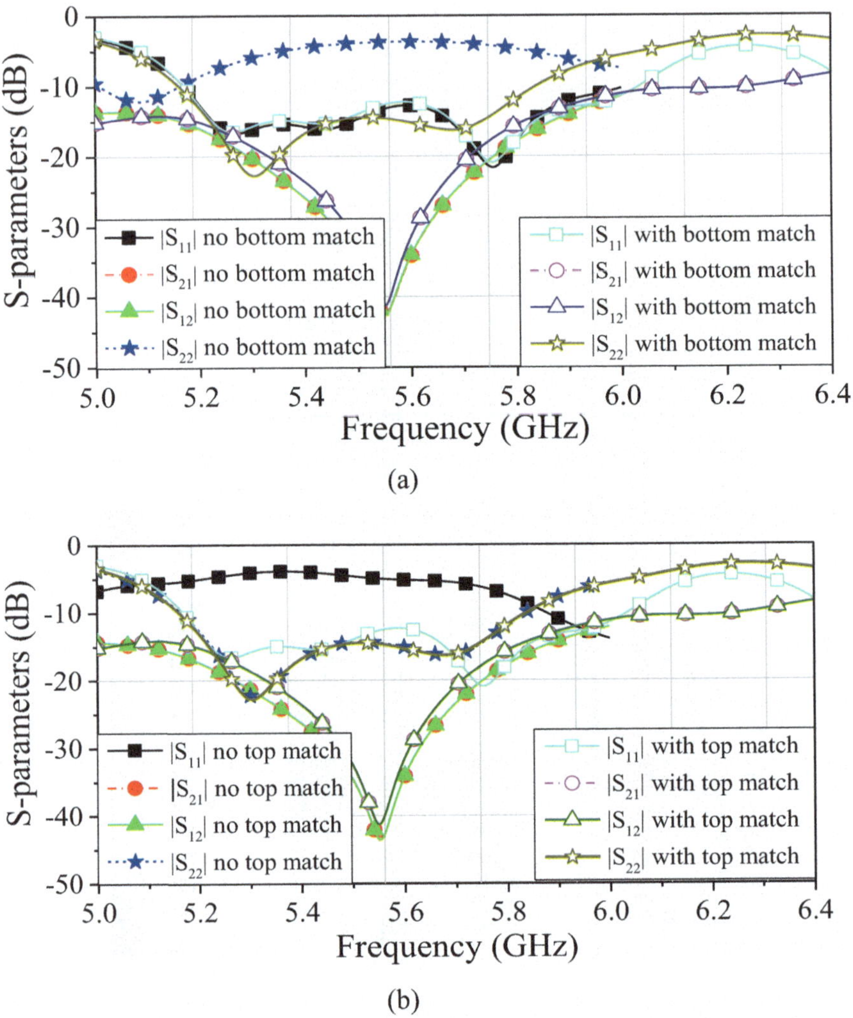

Fig. 5.6 Comparison of the S parameters between the cases with and without matching rings. **a** Bottom matching rings case. **b** Top matching rings case. Figure reproduced with permission from: [29], © 2019 IEEE

poor if the right match rings were removed, especially, |S22| will be higher than −10 dB during 5.2–6 GHz. But other S parameters (|S11|, |S21| and |S12|) were barely changed. This means the bottom impedance match rings better the impedance match of port 2, so that much more EM signals enter port 2, and radiated out from the slots array along the outside coaxial line from bottom to top.

Similarly, as shown in Fig. 5.4c, there are impedance match rings close to the top end too. These impedance match rings were optimized by similar methods, which are very useful to improve the impedance bandwidth of port 1. The comparison between

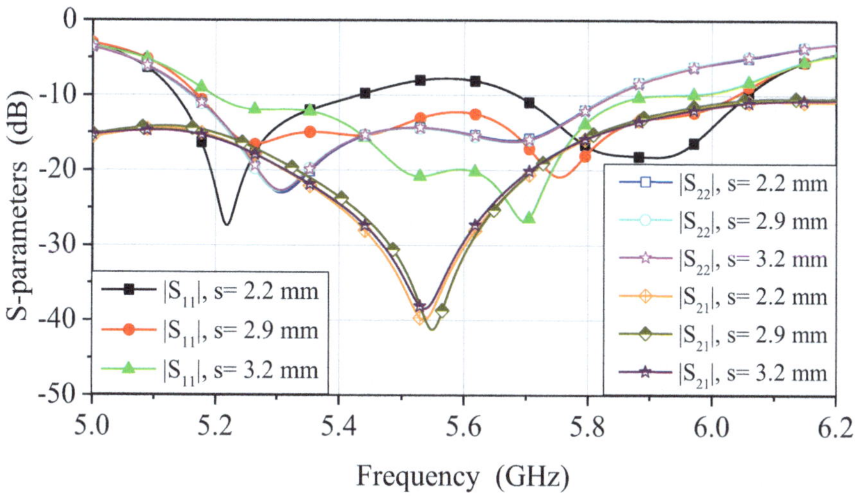

Fig. 5.7 Comparison of the S parameters: $|S_{11}|$, $|S_{22}|$ and $|S_{21}|$, between the cases with different slot widths, *s*, in the ring gap 's'. Figure reproduced with permission from: [29], © 2019 IEEE

the cases with the left impedance match rings or without the rings are shown in Fig. 5.6b, the |S11| becomes very bad if the left impedance match rings are removed, But the other S parameters (|S21|, |S12| and |S22|) almost remain unchanged. This means the right impedance rings better the impedance match of port 1, and much more EM signals entered port 1, and radiated out from the slots array along the outside coaxial line from top to bottom.

(ii) Impact of the width of the ring gap 's'

As shown in Fig. 5.4c, there is a thin ring gap between the top end and the outer metal of the inner coaxial wire, its width is *s*. The ring gap is just the throat for the EM signal from port 1 enter the radiation slots array. As shown in Fig. 5.7, the impedance bandwidth of port 1 is poor when the $s = 2.2$ mm, and the impedance bandwidth becomes better as $s = 2.7$ mm, but it becomes worse when *s* increases to 3.2 mm. Clearly, $|S_{11}|$ is very sensitive to the ring gap width.

On the contrary, $|S_{21}|$ and $|S_{22}|$ almost keep same when *s* vary from 2.2 mm to 3.2 mm as simulated in Fig. 5.7. It is that the *thin* ring gap is far away from the port 2 after the radiation slots, and the EM signal from port 2 firstly radiated out from the radiation slots array on the outside metal, then the little residual part will enter into the slot *s*.

5.3.4 Experimental Verification and Analysis

The optimized size parameters are list in Table 5.1, and the real improved DCPOA is manufactured according to the optimized size in Table 5.1, which is given in Fig. 5.8. The outside material of the antenna is copper, and the T-shaped connector is brass, both are connected by welding, and the connecting part are fixed by the copper sleeve. The real antenna is measured in anechoic chamber.

(i) S-parameters

Based on the optimized parameters values in Table 5.1 of the improved ODCPA, the simulated return loss and isolation curves are shown in Fig. 5.9. The available impedance bandwidth ($|S11| < -10$ dB) for port 1 of LHCP antenna is 0.86 GHz covering 5.17–6.03 GHz. But the impedance bandwidth ($|S22| < -10$ dB) for RHCP antenna is 0.677 GHz from 5.161 to 5.838 GHz. The isolation between these two ports is lower than -10 dB from 5.0 to 6.2 GHz, and the bandwidth of IL < -15 dB is 0.619 GHz during 5.197–5.816 GHz, the maximum isolation reaches 41 dB at $f = 5.55$ GHz, which means that both the RHCP antenna and LHCP antenna can work independently.

The measured return loss and isolation curves of the fabricated dual CP antenna with T-shaped connector are given in Fig. 5.9. The impedance bandwidth ($|S11| < -10$ dB) of LHCP antenna is 0.79 GHz, from 5.43 to 6.22 GHz. Comparing to the simulation results, the measured available band are shifted over about 200 MHz to higher frequency. Similarly, the measured impedance bandwidth with $|S22| < -10$ dB for RHCP antenna is 0.92 GHz over 5.48–6.4 GHz. The IL between these two antenna ports with ($|S21|, |S12| < -10$ dB) is from 5.0 to 6.2 GHz, which is same as the simulation results. And the bandwidth of IL < -15 dB is 0.57 GHz, from 5.49 to 6.06 GHz. It is clear that the frequency of the measured maximum isolation is moved to 5.785 GHz, about 235 MHz difference to the simulated result.

The differences between the experiment results and the simulated simulation results are mainly from the manufacture process. This improved omnidirectional dual CP antenna structure is very complicated as discussed in above, so that it is

Table 5.1 Optimized parameters of the improved DCPOA [29]

Parameter	Value (mm)	Parameter	Value (mm)	Parameter	Value (mm)
D_1	31.1	D_3	9.6	x_1	6.3
x_2	7.0	x_3	7.2	x_4	14.2
x_5	9.8	x_6	2.7	l_0	30.1
l_1	17	l_2	6.3	l_3	3.1
l_4	201.3	a_0	23.0	a_1	37.2
a_2	40.7	a_3	39.4	S	2.9
r_1	2.95	r_2	4.2	r_3	5.2
r_4	6.8	r_5	7	r_6	2.63

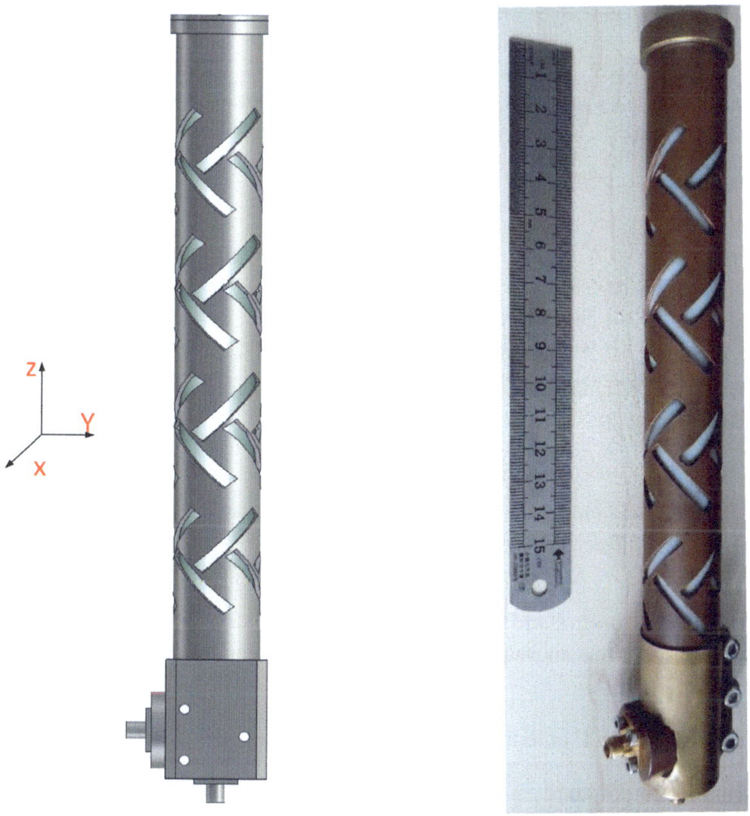

(a) (b)

Fig. 5.8 The optimized design of the dual circular polarization omnidirectional antenna. **a** Simulation model. **b** Fabricated prototype. Figure reproduced with permission from: [29], © 2019 IEEE

divided into several parts to be manufactured, such as the outside radiation slots array, the metal shell of the T-shaped connector, inner coaxial wire. Especially, the T-shaped feed structure is cut into two parts along the axis to manufacture, and then combined and squeezed together by the cooper sleeve, which leave minute gap, but may affect antenna work band. In another side, the radiation slot in the outside shell is dug out by the cuboid in the design model as shown in Fig. 5.8a, but it is fabricated by the cylindrical milling cutter, so that the corner of the slot is not right angle to decrease the girth of the slot. Thirdly, the thickness between two neighbor slots is varied for the manufacturing process to the corner of the slot, which is the bottle neck of the surface current. These fabrication errors are just the main reasons that generates the work bands difference between the simulation and the measurements. Finally, Teflon with relative permittivity to be $\varepsilon_r = 2.1$ is used to fill the cavity of the coaxial cylinder structure in the simulation models. But the real Teflon in the

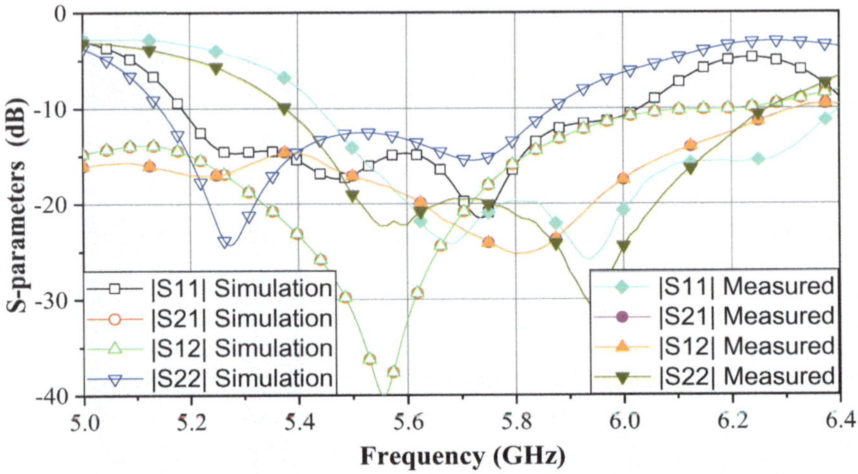

Fig. 5.9 The measured and the simulated $|S_{11}|$, $|S_{12}|$, $|S_{22}|$ results of the improved DCPOA with its T-shaped feed structure. Figure reproduced with permission from: [29], © 2019 IEEE

manufacturing is impure, and the actual value of its relative permittivity varies some around 2.1. This further increases the differences between the experiment results and the simulation results.

(ii) Far field of LHCP antenna

The normalized simulated far field patterns of the LHCP antenna excited from port 1 are shown in Fig. 5.10a–d. In the $\varphi = 0°$ plane, the far field vary with θcurves at f = 5.6 GHz, 5.8 GHz and 6 GHz are given respectively, in which the direction of main beams are almost at θ = 90°. And the 3 dB beam width are about 25°, which means the main beam is just like a round cake with omnidirectional radiation in the horizontal plane.

The normalized measured far fields of LHCP antenna excited from port 1 at $f =$ 5.6 GHz, 5.8 GHz and 6.0 GHz are shown in Fig. 5.11a–c. The main beam direction is almost at $\theta = 90^{o}$ at $f = 5.6$ GHz in Fig. 5.11a, and the cross polarization beams are smaller 15 dB than the main beam nearby $\theta = 90°$. The far field pattern in H-plane is close to a circle with small variation, which means this proposed LHCP antenna radiate omnidirectional in the horizontal, and consistent with the simulation.

In another side, the measured beams of LHCP antenna in Fig. 5.11 became fat near by $\theta = 110°–120°$, the reason is from the nonuniform radiation power distribution in the slots along the coaxial axis. As shown in Fig. 5.5a, the power flow distribution is much larger in the slots layers near the top of the antenna, but smaller in those slots close to the bottom.

Similarly, the normalized measured pattern at $f = 5.8$ GHz is in Fig. 5.11b, the main beam is almost at $\theta = 90°$, and it is omnidirectional radiation with little vibration in the H-plane. The cross polarization is smaller 18 dB than the main beam at $\theta =$

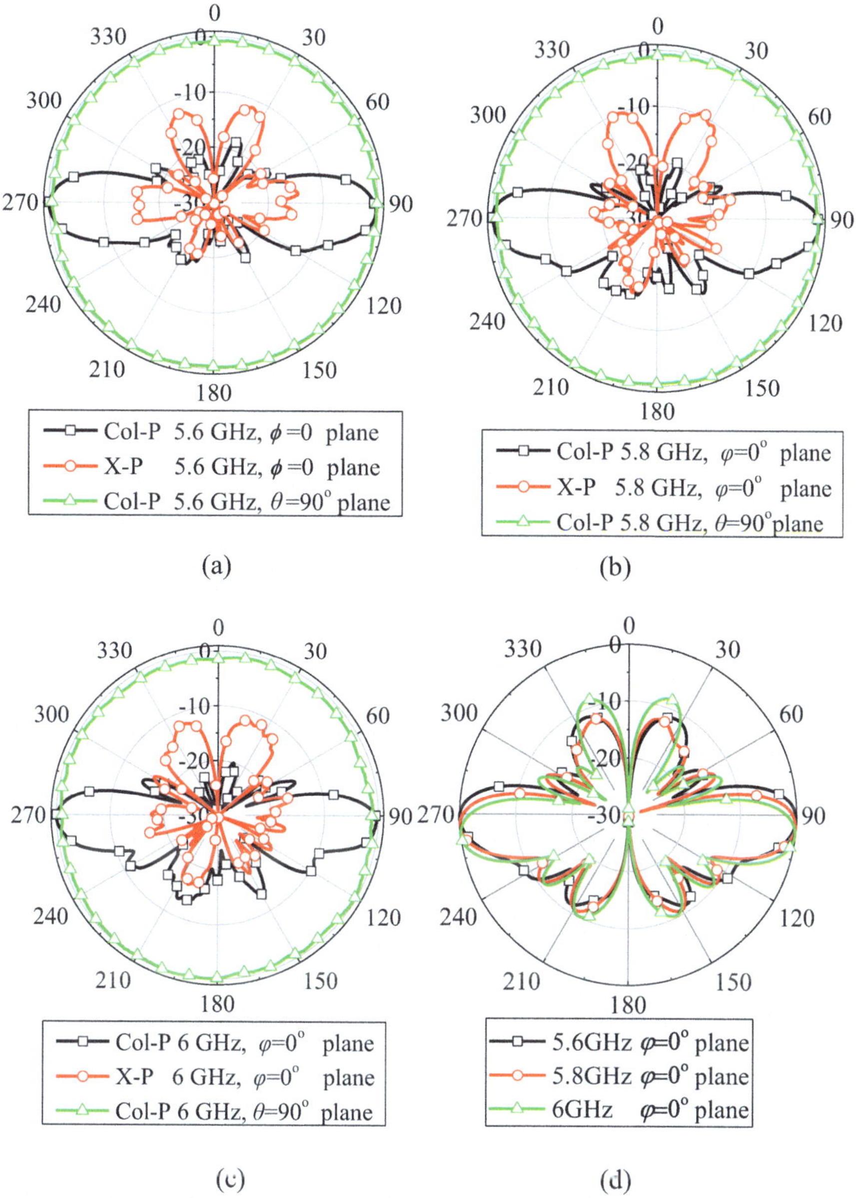

Fig. 5.10 Far-field patterns of the omnidirectional LHCP antenna of the TFS-enabled PDA. The normalized measured far-fields results at **a** $f = 5.6$ GHz, **b** $f = 5.8$ GHz, and **c** $f = 6.0$ GHz. **d** The corresponding normalized simulated co-polarization patterns in $\varphi = 0°$ plane. Figure reproduced with permission from: [29], © 2019 IEEE

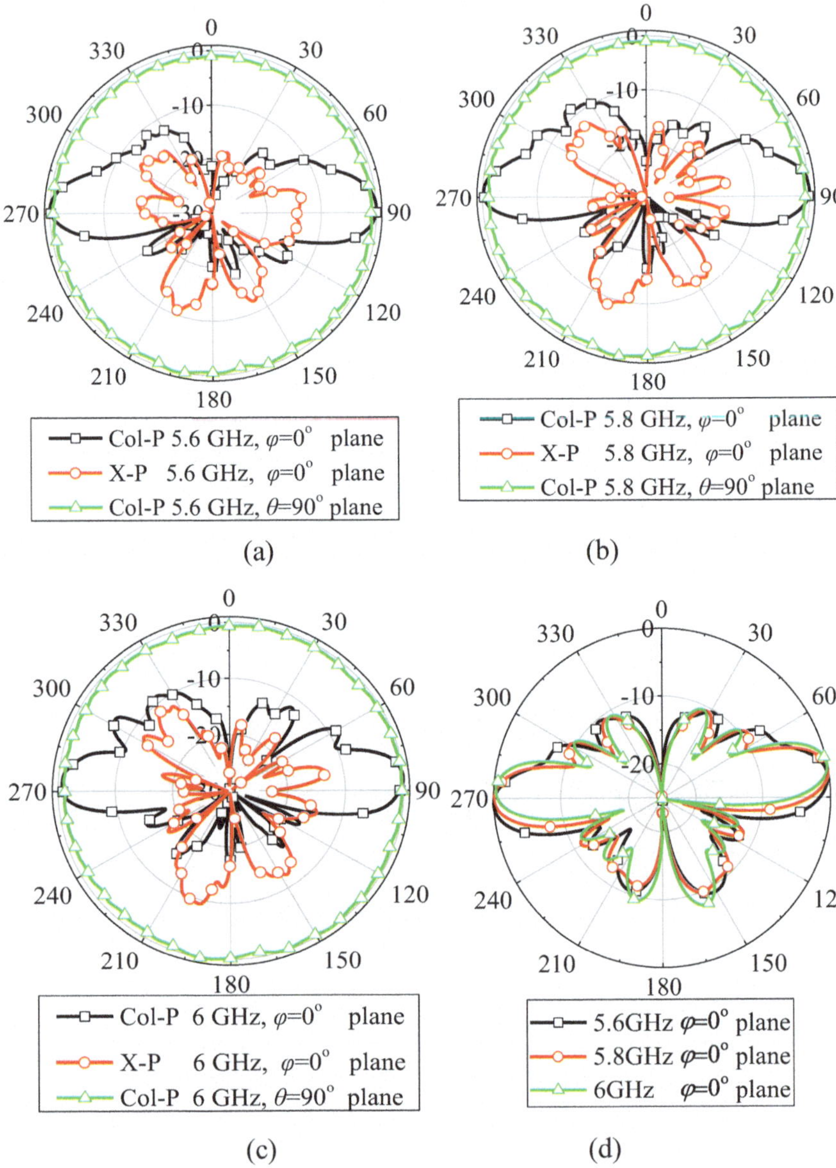

Fig. 5.11 Far-field patterns of the omnidirectional RHCP antenna of the TFS-enabled PDA. The normalized measured far-fields results at **a** $f = 5.6$ GHz, **b** $f = 5.8$ GHz, and **c** $f = 6.0$ GHz. **d** The corresponding normalized simulated co-polarization patterns in $\varphi = 0°$ plane. Figure reproduced with permission from: [29], © 2019 IEEE

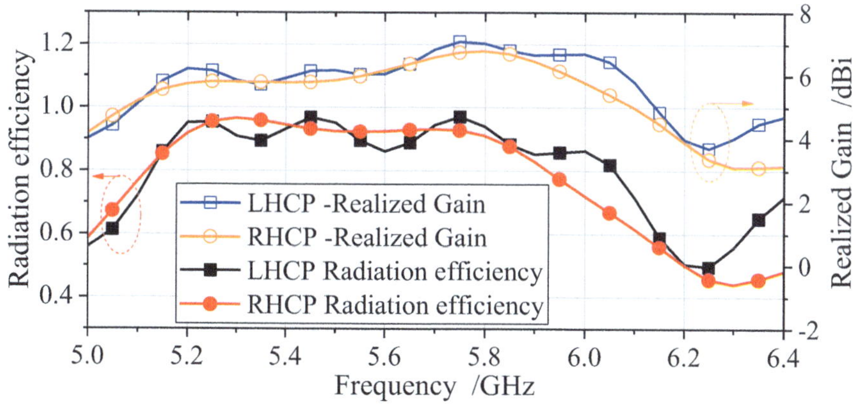

Fig. 5.12 Simulated radiation efficiency and realized gain values of the LRHCP and RHCP antennas as functions of the source frequency. Figure reproduced with permission from: [29], © 2019 IEEE

90°. The main beam at $f = 5.8$ GHz became fatter near by $\theta = 110°–120°$ than that at $f = 5.6$ GHz.

The normalized measured pattern at $f = 6.0$ GHz is in Fig. 5.11c, the main beam is almost at $\theta = 90°$, and it is omnidirectional radiation with little vibration in the H-plane. The cross polarization is smaller 17.5 dB than the main beam at $\theta = 90°$. There is the 1st side beam at $f = 6.0$ GHz. In one side, the inhomogeneous power flow distributions such as that in Fig. 5.5a result in the unsymmetrical beam in E-plane. In another side, the electrical interval length between the slot rounds become larger with the frequency increasing from 5.6 to 6.0 GHz, so that a new sink is introduced at $\theta = 110°$, and the main beam is split.

The measured far-field patterns of the LHCP antenna in the $\varphi = 0°$ plane at 5.6, 5.8 and 6.0 GHz are shown in Fig. 5.11a–c. They are similar as the simulated results in Fig. 5.11d.

(iii) Far field of the RHCP antenna

The far field patterns of RHCP antenna excited from port 2 are similar to the LHCP results too, which are given in Fig. 5.11a–d. The far field patterns in E-plane ($\varphi = 0°$) at $f = 5.6$ GHz, 5.8 GHz and 6.0 GHz are almost symmetrical respectively. The direction of main beams are very close to $\theta = 90°$, the 3 dB beam width are about 24°. Correspondingly, the far field in H-plane ($\theta = 0°$) is omnidirectional. The normalized measured far fields of RHCP antenna excited from port 2 at $f = 5.6$ GHz, 5.8 GHZ and 6.0 GHz are presented in Fig. 5.11. In E-plane, $f = 5.6$ GHz, the main beam is in the horizontal ($\theta = 90°$), and the cross polarization beams is smaller about 15 dB than the main beam at $\theta = 90°$ in Fig. 5.11a. The main beam in H-plane is omnidirectional with little vibration, but consistent with the simulation.

Similarly, the normalized measured pattern of RHCP antenna at $f = 5.8$ GHz is in Fig. 5.11b, the main beam direction in E-plane is almost at $\theta = 90°$, and it is omnidirectional radiation with little vibration in the H-plane.

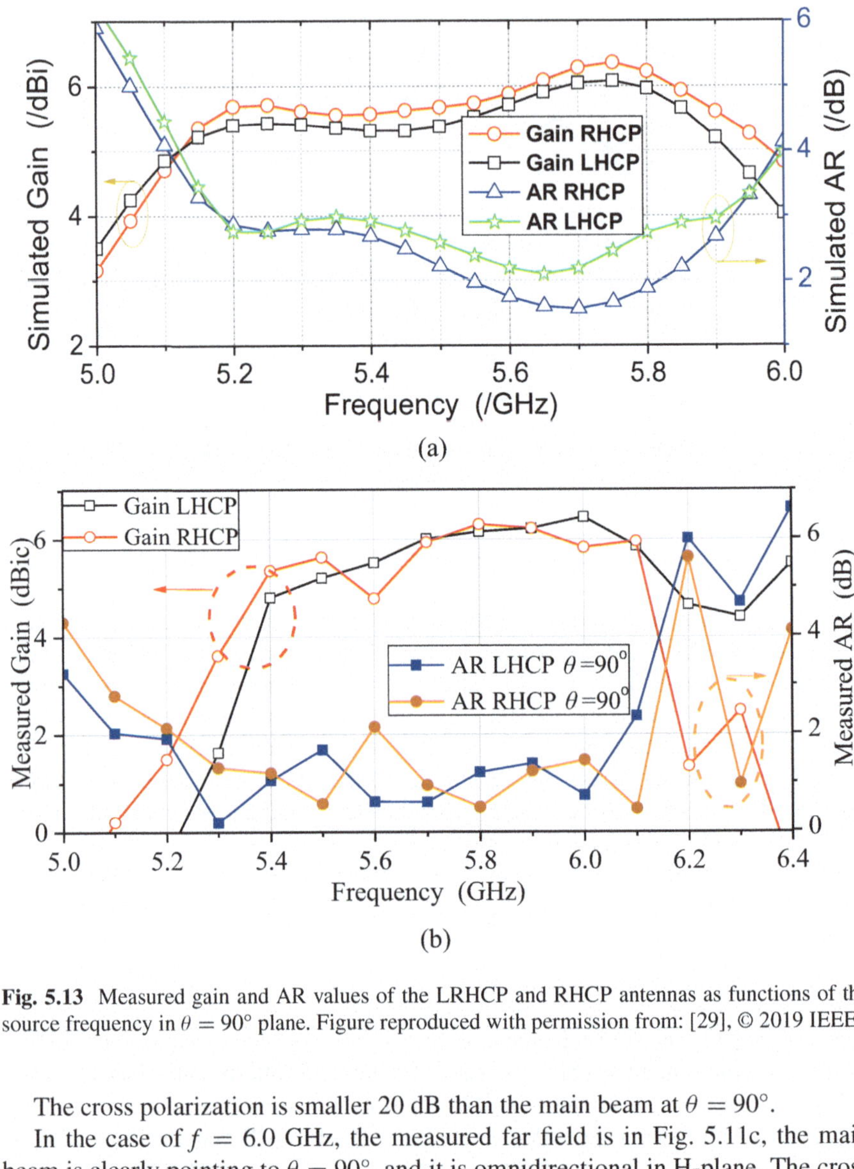

Fig. 5.13 Measured gain and AR values of the LRHCP and RHCP antennas as functions of the source frequency in $\theta = 90°$ plane. Figure reproduced with permission from: [29], © 2019 IEEE

The cross polarization is smaller 20 dB than the main beam at $\theta = 90°$.

In the case of $f = 6.0$ GHz, the measured far field is in Fig. 5.11c, the main beam is clearly pointing to $\theta = 90°$, and it is omnidirectional in H-plane. The cross polarization beam at $\theta = 90°$ is lower 20 dB than the main beam.

The main beams in Fig. 5.11 all are almost asymmetrical between the upper and lower parts. In Fig. 5.11a, $f = 5.6$ GHz, the pattern in E-plane is asymmetrical between the upper and lower parts of the main beam, the upper is fatter than the lower. In Fig. 5.11b, $f = 5.8$ GHz, The main beam has slightly split near by $\theta = 60°$ in E-plane. In Fig. 5.11c, $f = 6.0$ GHz, the main beam has clearly split around $\theta = 60°$. The reason is similar as we discussed in above, as shown in Fig. 5.11b, the power flow is nonuniform along the axis, and much larger in the lower 3 round slots. In another side, the electrical interval distance become larger with frequency increasing from 5.6 to 6.0 GHz, so that the 1st sub beam was split out.

The measured far-field patterns of the RHCP antenna in the = 0° plane at 5.6, 5.8 and 6.0 GHz are shown in Fig. 5.11a–c. They are similar to the simulated ones given in Fig. 5.11d.

(iv) Radiation performance of the improved DCPOA

In Fig. 5.11 the simulated radiation efficiencies and realized gains of the designed dual-CP antenna of LHCP or RHCP are presented. For LHCP antenna, the radiation efficiency is larger than 0.8 during 5.13–6.05 GHz, and the realized gain is 5.0–7.0 dBi including the return loss Fig. 5.11. Similarly, the simulated available impedance band of the RHCP antenna is from 5.11–5.9 GHz and the corresponding radiation efficiencies is larger than 0.8. The simulated realized gain is 5.0–7.0 dBi considering the return loss in Fig. 5.11.

The simulated gain and axial ratio of the dual CP antenna are given in Fig. 5.11a. For the LHCP antenna (excited from port 1), the CP band is about 650 MHz from 5.2 to 5.85 GHz, and corresponding gain is 5.3–6.07 dBc in this band. For the RHCP antenna (excited from port 2), the CP band is about 710 MHz from 5.2 to 5.91 GHz, and corresponding gain is 5.5–6.4 dBc. It is noted that the RHCP antenna performs better than the LHCP antenna in simulation.

The measured gain and axial ratio of the dual CP antenna are shown in Fig. 5.11b. For the LHCP antenna, the LHCP band of AR < 3 dB is from 5.05–6.34 GHz except for 6.19–6.22 GHz, which is similar as the simulation in Fig. 5.11a. The measured gain of the LHCP antenna is 2.5–9.5 dB in the CP band, and the maximum gain is at 6.2 GHz. The measured gain curve is similar as the simulation in Fig. 5.11a too, but the main CP band shift 200 MHz to higher frequency.

For the RHCP antenna (excited from port 2), the measured CP band of AR < 3 dB is from 5.1 to 6.36 GHz except for 6.15–6.25 GHz, and corresponding gain is 4–9dBc, and the gain peak is at 6.2 GHz. The measured gain curve is similar as the simulation in Fig. 5.11a, the main CP band shift 200 MHz to higher frequency too.

The measured main LHCP and RHCP bands are shifted about 250 MHz to higher frequencies from their simulated values. Similar as noted above, the reasons are mainly due to the various fabrication and filling medium errors (Fig. 5.12).

5.3.5 *Comparison with Other Dual CP Antenna*

The comparison between the improved DCPOA prototype and other dual CP antennas are listed in Table 5.2. The measured impedance bandwidth of the improved DCPOA is 16.4% for its LHCP antenna and 15.28% for its RHCP antenna. The isolation of the improved DCPOA prototype between the two feeding ports is 25 to 10 dB during their work bands. The measured gain and AR values of both LHCP and RHCP

Table 5.2 Comparison of dual CP omnidirectional antenna [29]

Ref.	CP		BWs (%)	CP band (%)	Omni-direction	size	Gain (dBic)
[12]	CP/switch		19.8	20	Good	0.93λ × 0.93λ × 0.024λ	< 2.5
[24]	CP/switch		18	22	Good	0.55λ × 0.44λ × 0.44λ	0.1–0.4
[25]	CP		41	45	Good	1.625λ × 0.38λ × 0.38λ	1.5–4.5
[15]	DCP	LHCP	16.4	18.1	Poor	3.5λ × 0.6λ × 0.6λ	4–6
		RHCP	16.4	18.1	Poor	3.5λ × 0.6λ × 0.6λ	4–6
This work	DCP	LHCP	16.4	21.35	Good	4.3λ × 0.6λ 0.6λ	5–6.46
		RHCP	15.28	20.58	Good	4.3λ × 0.6λ × 0.6λ	4.8–6.32

antennas from Fig. 5.13 show that the peak LHCP gain is 5.0–6.46 dBic while 4.8–6.32 dBic gain for the RHCP antenna over their work bands. So, the improved DCPOA prototype antenna has very good dual CP omnidirectional character in the azimuthal plane.

5.4 TFS-Enabled Space Diversity Antenna

With the development of air base station, such as balloon base station, it will cover wider region and transmit less attenuation to the user on earth than the traditional base station on the ground. It is necessary that the antenna of terminal on the ground should achieve hemisphere coverage to connect the base station in the sky and the ground base station in the network. Space diversity antenna is a good solution method. Space diversity antenna radiates different far field patterns from different ports, which is good to be against the dark fringes of coverage, multipath and channel fading of the wireless communication.

5.4.1 The Principle of the SDA with TFS

The spatial diversity antenna is actually realized by putting two antennas together to achieve hemisphere coverage. The omnidirectional CP slots array antenna (OCPSA) on the bottom achieves high gain in the horizontal plane, and a low profile directional planar equiangular spiral antenna (LPDPESA) is mounted on the top, which cover the top space [21]. The TFS is used to feed these two antennas to obtain the space diversity antenna (SDA) as shown in Fig. 5.14. The ITCC runs through the top cap of the coaxial slot array antenna without the ring gap 's'. It just like the port 3 being extended, which can feed a low-profile antenna on the top. So, port 1 directly feeds the LPDPESA on the top to directional radiation pattern, and port 2 feeds the OCPSA from the TFS to omnidirectional radiation pattern. The organic combination of TFS, OCPSA, and low-profile DPESA was optimized to the desired performance.

5.4.2 SDA Structure and Performance

The TFS, the OCPSA and the LPDPESA are combined to construct the SDA, as shown in Fig. 5.14. The ITCC with characteristic impedance 50 Ω completely passes through the center of the OCPSA, and straightly connected with port 1 and the LPDPESA on the top. Here, the LPDPESA is an optimized planar directional antenna with low profile with 50 Ω SMA adaptor port, as reported in [21]. The radiation part of the low-profile directional antenna is equiangular planar spiral structure, and there are two parasitic layers with rotational symmetry structures—optimized digital-code metasurface—under the radiation part, and there is a rectangular metal bottom.

Furthermore, the metal shell of the ITCC is regarded as the center of the OCPSA fed by port 2. So, the feeding channel from port 1 to the LPDPESA and the feeding channel from port 2 to the OCPSA are physically isolated.

When the EM signal is fed from port 2, it runs through port 4 of the TFS and then enters into the OCPSA from bottom to top, which is similar as the case of the RHCP antenna of the IDCPOA in Sect. 5.3, where the omnidirectional RHCP EM field is radiated out. Similarly, the bottom impedance match rings are loaded to improve the impedance bandwidth of port 2. The bottom impedance match rings are optimized based on the transmission line theory to get good return loss. In addition, the distance of the fourth circle of the slot pairs to the top metal cap is also adjusted for good impedance matching. The space between the outer conductor of the ITCC and the outer conductor shell of the antenna is filled with Teflon, which has ε_r of 2.1. The proposed antenna is simulated and optimized by CST and the optimized parameters are shown in Table 5.3.

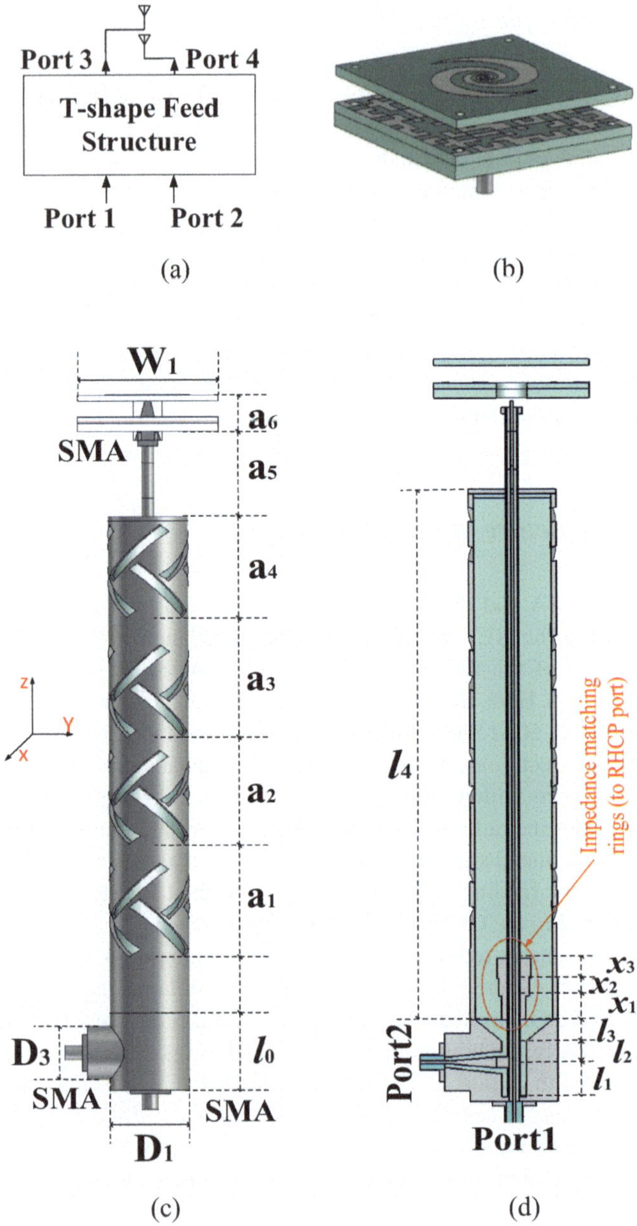

Fig. 5.14 The T-shaped feed CP slots antenna with embedded coaxial line to feed the directional planar equiangular spiral antenna on the top. **a** Block diagram, **b** The low-profile, wide bandwidth, directional planar equiangular spiral antenna (DPESA) on top of the structure has CP characteristics [21], **c** 3D view. **d** Cross-sectional view through a vertical cut. Figure reproduced with permission from: [18, 29], © 2019 IEEE

Table 5.3 Optimized design parameters of the spatial diversity antenna

Parameter	Value (mm)	Parameter	Value (mm)	Parameter	Value (mm)
D_1	25	D_3	19	a_1	39.7
a_2	37.5	a_3	42.25	a_4	35.4
a_5	30	a_6	12.85	l_0	27
l_1	11.5	l_2	7	l_3	7
l_4	174.6	x_1	7.8	x_2	5.8
x_3	6.5	W_1	44.2		
r_1	2.9	r_2	4.0	r_3	5.2

5.4.3 Experimental Verification and Analysis

The SDA prototype is manufactured as shown in Fig. 5.15. Its impedance bandwidth is measured by Agilent PNA Network Analyzer E8361C, and the radiation performances are measured in the far field anechoic chamber.

(i) S-parameters

The measured return loss and isolation curves of the fabricated SDA are shown in Fig. 5.16. The bottom OCPSA achieves the impedance bandwidth of 730 MHz from 5.52 to 6.35 GHz, with a bandwidth of about 10.9%. The measured impedance bandwidth of the LPDPESA on top is wider than 1500 MHz from 5.0 to 6.5 GHz. The measured isolation between the two ports is lower than −30 dB from 5.0 to 6.5 GHz, which can be seen in Fig. 5.16. It is noted that the measured impedance band RHCP omnidirectional antenna shifts about 380 MHz to higher frequency than the simulated results of the RHCP omnidirectional antenna of the IDCPOA in Fig. 5.8 in Sect. 5.3. The frequency offset is due to the manufacturing error of the TFS, slots array on the shell, size error between the slots and the filled impure Teflon in the antenna.

Again, it has been determined through simulation and further measurements that this frequency offset is due to the manufacturing errors of the TFS, the slot array on the outer shell, the distance of separation errors between the slots, and the real nature of the Teflon filling.

(ii) SDA far-field performance

The normalized measured far fields at $f = 5.6$ GHz, 5.8 GHz and 6 GHz of the LPDPESA on the top are given in Fig. 5.17a–c, respectively. The main beams point upward, and the 3 dB beam width in $\varphi = 0°$ plane are 66.5°, 72° and 69°, respectively. The LPDPESA is from the equiangular planar spiral structure, which is a UWB structure with bidirectional radiation. To achieve the desired wideband directional radiation with low profile structure, two parasitic layers, the digital-code metasurface [22, 23] is introduced under the equiangular planar spiral structure to achieve directional radiation in wide band. Particle swarm optimization is employed for generating

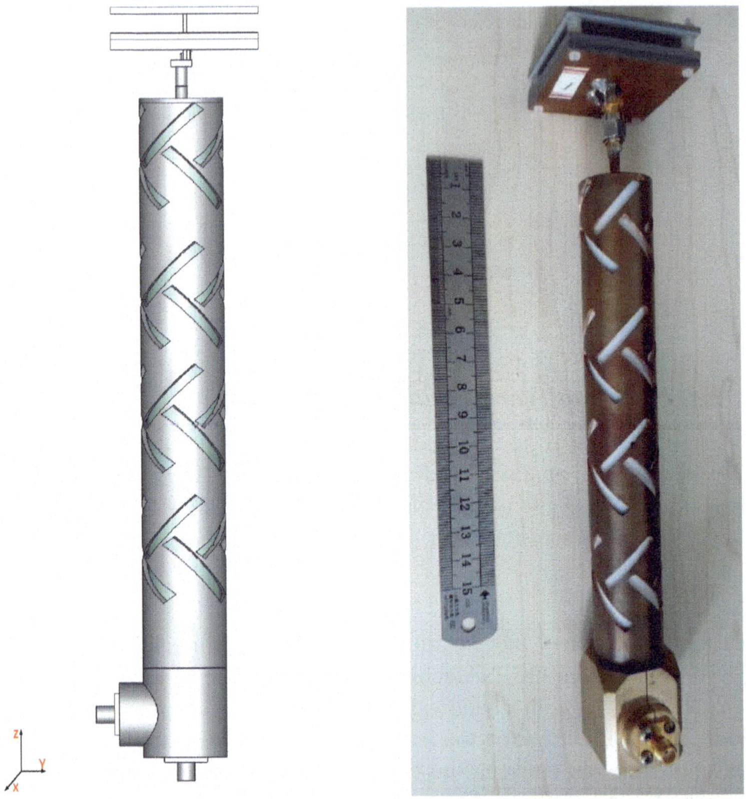

(a) (b)

Fig. 5.15 The optimized CP SDA. **a** Simulation model. **b** Fabricated prototype. Figure reproduced with permission from: [29], © 2019 IEEE

metamaterial parasitic layers. By inserting such metamaterial layers, a low profile wideband CP antenna with directional radiation pattern and high gain was designed. There also is metal bottom being as ground to DPESA. The parasitic digital-code metasurface layers and the ground have been stacked to reflect the downward fields to upward radiation from these structures to achieve the correct phase to mitigate the back radiation. Unfortunately, the trade-off with this stacking arrangement is a degradation of the CP performance.

The normalized measured far fields patterns of the omnidirectional RHCP antenna excited from port 2 are shown in Fig. 5.18a–c. The main beam in θ-plane at $f =$ 5.6 GHz is in the horizontal ($\theta = 90°$), and the cross-polarization beams is about 18 dB smaller than the main beam at $\theta = 90°$ in Fig. 5.18a. The far field in φ-plane ($\theta = 90°$) is omnidirectional with little vibration. At $f = 5.8$ GHz, the main beam in θ-plane is in the horizontal ($\theta = 90°$) too, and the cross-polarization beams is smaller

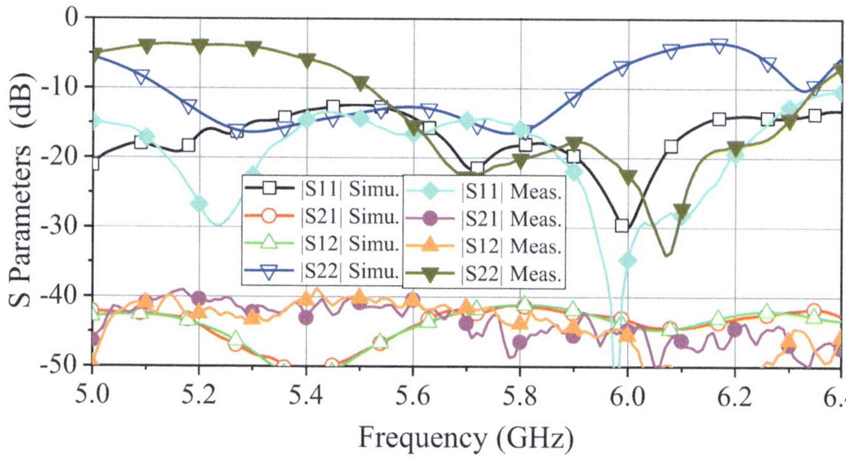

Fig. 5.16 The measured and simulated |S11|, |S12|, |S21|, |S22| results of the TFS-enabled SDA. Figure reproduced with permission from: [29], © 2019 IEEE

about 20 dB than the main beam at $\theta = 90°$ in Fig. 5.18b. The far field in φ-plane ($\theta = 90°$) is omnidirectional with little vibration. In the case of $f = 6.0$ GHz, the measured far field is shown in Fig. 5.18c, the main beam clearly points to $\theta = 90°$, and it is omnidirectional in φ-plane ($\theta = 90°$). The cross polarization beam at $\theta = 90°$ is 15 dB lower than the main beam. The measured far field patterns coincide with the results of the RHCP omnidirectional antenna of the IDCPOA in Sect. 5.3.

(iii) SDA radiation performance

In Fig. 5.18, the simulation results of the radiation efficiency curve and realized gain curve of the top LPDPESA (excited from port 1) and the omnidirectional RHCP antenna (excited from port 2) are presented. For the LPDPESA, the simulated radiation efficiency is larger than 0.9 during 5.0–6.4 GHz. And the frequency band is same as the simulated impedance band with $|S_{11}| < -10$ dB and $|S_{22}| < -10$ dB in Fig. 5.18. For the omnidirectional RHCP antenna, the radiation efficiency is larger than 0.8 in 5.05–6.0 GHz with the corresponding simulated realized gain 3.63–7.82 dBi.

The simulated gain and axial ratio of the spatial diversity antenna are given in Fig. 5.18a. For the LPDPESA (excited from port 1) on the top, the simulated gain is larger than 7 dBi from 5.2 to 6.0 GHz. For the RHCP antenna (excited from port 2), the CP band is about 720 MHz from 5.1 to 5.82 GHz, and corresponding gain is 5.2–6.8 dBic. It is noted that the RHCP antenna is almost the same as the case of the RHCP omnidirectional antenna of the dual circular polarization antenna in Sect. 5.3.

The measured gain and axial ratio of the SDA are shown in Fig. 5.18b. For the LPDPESA (excited from port 1) on the top, the measured gain is 4–7.8 dBic in

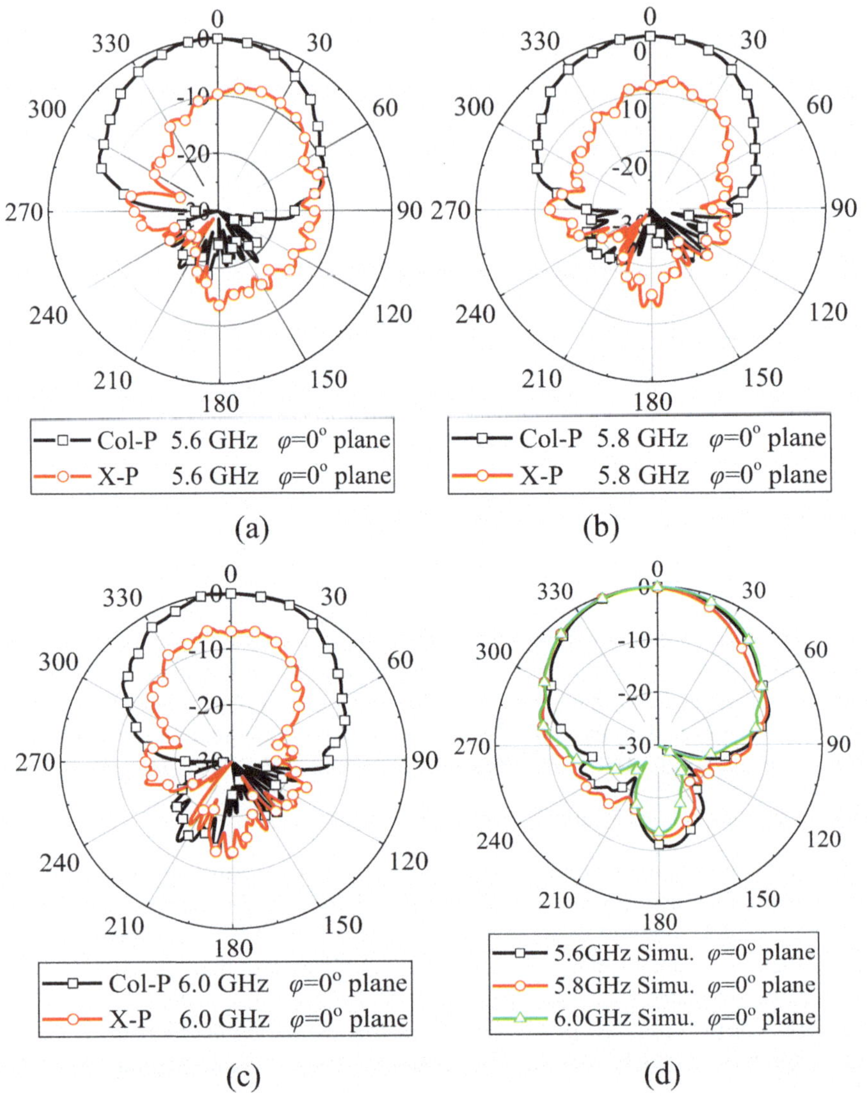

Fig. 5.17 The normalized measured far field patterns of the DPESA of the TFS-enabled SDA at **a** $f = 5.6$ GHz, **b** $f = 5.8$ GHz, and **c** $f = 6.0$ GHz. **d** The corresponding normalized simulated co-polarization patterns of the DPESA in the $\varphi = 0°$ plane. Figure reproduced with permission from: [29], © 2019 IEEE

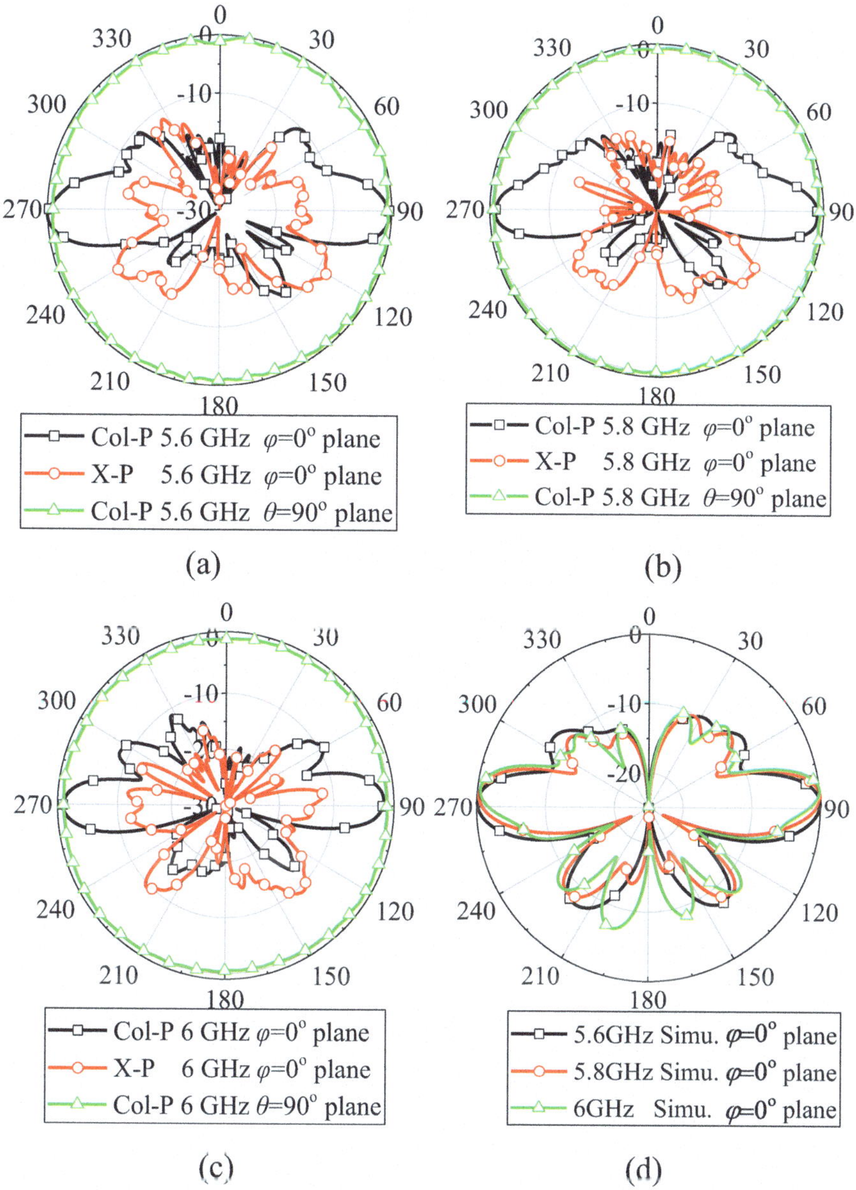

Fig. 5.18 Far-field patterns of the omnidirectional RHCP antenna of the TFS-enabled SDA. The normalized measured far-fields results at **a** $f = 5.6$ GHz, **b** $f = 5.8$ GHz, and **c** $f = 6.0$ GHz. **d** The corresponding normalized simulated co-polarization patterns in $\varphi = 0°$ plane. Figure reproduced with permission from: [29], © 2019 IEEE

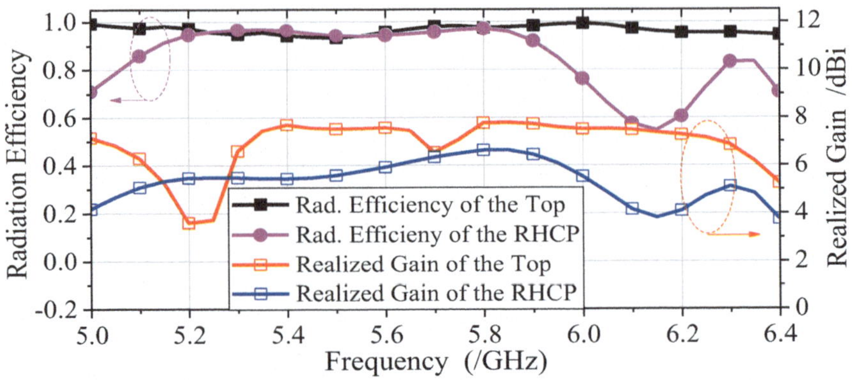

Fig. 5.19 The simulated radiation efficiency and the realized gain of the SDA antenna's omnidirectional RHCP system and its low-profile DPESA. Figure reproduced with permission from: [29], © 2019 IEEE

5.4–6.4 GHz, and the maximum gain is at 5.6 GHz. There is about 450 MHz (5.45–5.9 GHz) CP band of $AR < 3$ dB in the upward direction for the measured AR curve.

For the RHCP antenna (excited from port 2), the measured CP band is about 816 MHz from 5.3 to 6.16 GHz, and the corresponding gain is 2–7 dBic, and the maximum gain is at 6.1 GHz. Comprehensively considering the impedance band in Fig. 5.18, and RHCP band, the available working band is 5.5–6.16 GHz, and the gain is 4.4–6.99 dBic in this available band. This is almost the same as the case of the RHCP omnidirectional antenna of the dual circular polarization antenna in Sect. 5.3 (Fig. 5.19 and 5.20).

5.4.4 Comparison with Other Spatial Diversity Antennas

The comparison of the proposed spatial diversity antenna and other diversity antennas are given in Table 5.4. The proposed spatial diversity antenna achieves the impedance bandwidth of 10.9% in experiments. Both S_{11} and S_{22} are lower than −10 dB between 5.52 and 6.35 GHz and the isolation between these two ports is lower than −30 dB within the same band, which can be seen in the Fig. 5.16. The gain and AR of the antenna are shown in Fig. 5.16. The measured gain of the low profile directional antenna (excited from port 1) on the top is 4–7.8 dBic in the work band. The measured CP band of the RHCP antenna (excited from port 2) is about 816 MHz, and the measured gain is 2–7 dBic. Larger bandwidth and high gain levels for both directional and omniditectional radiations are organically combined together, so that it is very suited for the hemispherical coverage with spatial diversity to a wide variety of wireless applications.

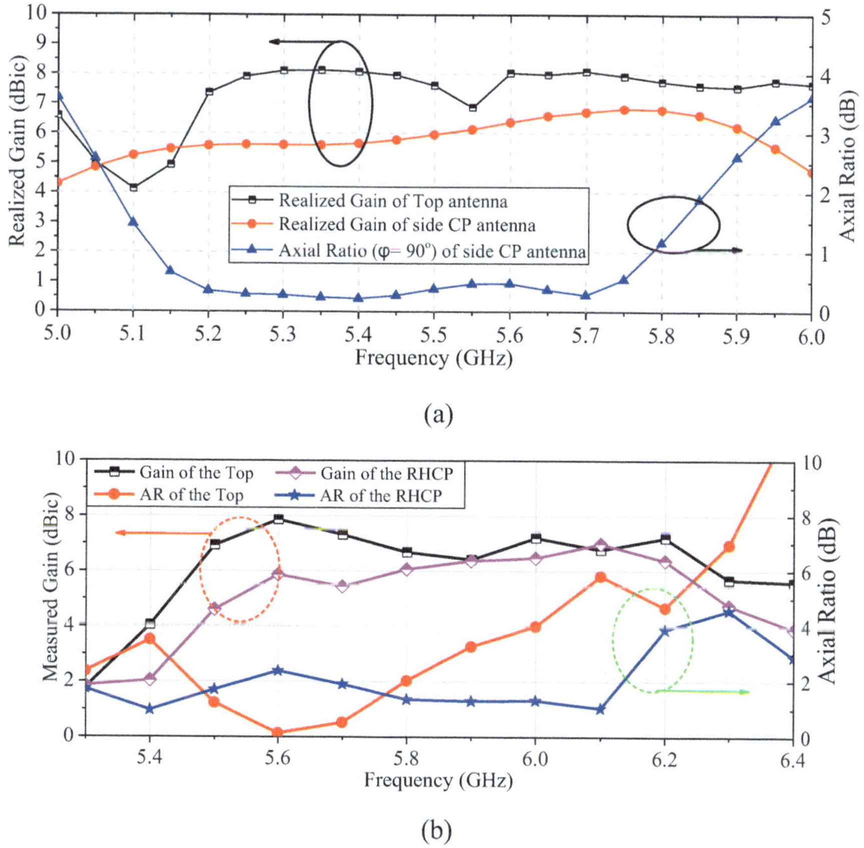

Fig. 5.20 The measured gain and AR values of the SDA antenna's omnidirectional RHCP system in $\theta = 90°$ plane and its low-profile DPESA in the upward direction. Figure reproduced with permission from: [29], © 2019 IEEE

Table 5.4 Comparison of SDA [29]

Ref.	Bands (GHz)	Polarization	Pattern type	Size (mm^3)	Gain (dBi)
[26]	1.7–2.7 4.7–8.5	LP	Directional	50 × 17 × 0.8	< −0.313
[27]	27.56–28.4 37.8–38.9	RHCP	Directional	3 × 4 × 0.254	7.2
[28]	4.6% 0.85–0.89	LHCP	End fire	229.2 × 26.2 × 26.2	2.5
This work	14.35% 5.5–6.35	RHCP	Omni-directional	243 × 44 × 44	< 6.99
	Wide band	Top: RHCP	Directional		< 7.8

5.5 Summary

In this chapter, based on the proposed compact 4-port TFS, the OCPSA and low profile directional antenna are combined to construct the polarization diversity antenna and spatial diversity antenna. Both diversity antennas are fed from the down part, which is convenient to mount and use.

The proposed IDCPOA achieves the impedance bandwidth of 12.7% in experiments. Both S_{11} and S_{22} are lower than -10 dB between 5.45 and 6.22 GHz, and the isolation between the two ports is lower than -15 dB within the same band. The LHCP gain is 6.88–9.06 dBic while the average RHCP gain is 6.54–9.6 dBic. The manufactured antennas achieves good CP properties in the horizontal omnidirection. It can be found that the proposed antenna performs a good omnidirectional character. The measured impedance bandwidth of the proposed SDA is about 10.9%, range from 5.52 to 6.35 GHz and the isolation between the two ports is lower than -30 dB within the same band. The gain of the LPDPESA is 4–7.8 dBic in the work band, and the main beam points upward. The CP band of the RHCP antenna (excited from port 2) is about 816 MHz, and the gain is 4.4–6.99 dBic in the comprehensively available band. The RHCP antenna performs good horizontal omnidirectional CP radiation.

It is noted that the TFS and the OCPSA can be conveniently constructed the circular polarization antenna and the diversity antenna, the PDA and SDA systems make both of them attractive to a variety of applications, which are very useful to rise the performance of the wireless communication.

References

1. Y. Kihc, M. Koca, E. Ananm, Space-time-polarization diversity in multiple-input multiple-output communication systems, in *IEEE AFRICON 2009*, Nairobi, Kenya, Sept. 23–25, 2009
2. W. Lee, Y. Yu, Polarization diversity system for mobile radio. IEEE Trans. Commun. **20**(5), 912–923 (1972)
3. R.U. Nabar, H. Bölcskei, V. Erceg, D. Gesbert, A.J. Paulraj, Performance of multiantenna signaling techniques in the presence of polarization diversity. IEEE Trans. Signal Process. **50**(10), 2553–2562 (2002)
4. G. Maral, M. Bousquet, *Satellite Communication Systems, Chichester* (Willey, England, 2002)
5. M. Richharia, *Satellite Communication Systems* (McGraw Hill, New York, 1999)
6. G. Chenhu, J. Geng, H. Zhou, J. Li, L. Liu, Y. Chen, Y. Liang, X. Liang, W. Zhu, R. Jin, A half-space covered antenna for air-ground communication, in *IEEE Asia-Pacifc Conference on Antennas and Propagation*, Xi'an, China, Oct. 16–19, 2017
7. E. Fishler, A. Haimovich, R.S. Blum, D. Chizhik, R.A. Valenzuela, Spatial diversity in radars-models and detection performance. IEEE Trans. Signal Process. **54**(3), 823–838 (2006)
8. A.M. Hunter, J. Andrews, S. Weber, Transmission capacity of ad hoc networks with spatial diversity. IEEE Trans. Wireless Commun. **7**(12), 5058–5071 (2008)
9. S.N. Diggavi, N. Al-Dhahir, A. Stamoulis, A.R. Calderbank, Great expectations: the value of spatial diversity in wireless networks. Proc. IEEE **92**(2), 219–270 (2004)
10. B. Li, S.W. Liao, Q. Xue, Omnidirectional circularly polarized antenna combining monopole and loop radiators. IEEE Antennas Wirel. Propag. Lett. **12**, 607–610 (2013)

11. X.-L. Quan, R.L. Li. Broadband dual-polarized omnidirectional antennas, in *2012 IEEE Antennas and Propagation Society International Symposium (APSURSI)*, vol. 11, no. 4, Chicago, IL, USA, July, 8–14, 2012, pp. 1–2
12. Y.M. Cai, S. Gao, Y. Yin, W. Li, Q. Luo, Compact-size low-profile wideband circularly polarized omnidirectional patch antenna with reconfigurable polarizations. IEEE Trans. Antennas Propag. **64**(5), 2016–2021 (2016)
13. B. Zhou, J. Geng, X. Bai, L. Duan, X. Liang, R. Jin, An omnidirectional circularly polarized slot array antenna with high gain in a wide bandwidth. IEEE Antennas Wirel. Propag. Lett. **14**, 666–669 (2015)
14. B. Zhou, J. Geng, X. Liang, R. Jin, A compact omnidirectional CP coaxial slots antenna, in *2015 IEEE 6th International Symposium on Microwave, Antenna, Propagation, and EMC Technologies (MAPE)*, Shanghai, China, Oct. 28–30, 2015, pp. 821–822
15. B. Zhou, J. Geng, Z. Li, W. Wang, X. Liang, R. Jin, Dual circularly polarized omnidirectional antenna with slot array on coaxial cylinder. Int. J. Antenna Propag. 1–7 (2015), Article ID 127820. https://doi.org/10.1155/2015/127820
16. B. Zhou, J. Geng, X. Liang, R. Jin, G. Chenhu, Omnidirectional circularly polarized antenna with high gain in wide bandwidth, in *Modern Antenna Systems* (Intech, 2016). ISBN: 978-953-51-4943-9
17. G. Chenhu, J. Geng, L. Liu, H. Zhou, X. Zhao, Y. Liang, X. Liang, W. Zhu, R. Jin, A circular truncated cone slot antenna with circular polarized conical beam, in *2017 IEEE International Symposium on Antennas and Propagation and USNC/URSI National Radio Science Meeting*, San Diego, CA, USA, July 9–14, 2017, pp. 1533–1534
18. G. Chenhu, J. Geng, B. Zhou, J. Li, Y. Chen, W. Zhu, R. Jin, R.W. Ziolkowski, A T-shaped feed structure to enhance the performance of a polarization diversity antenna, in *2017 IEEE International Symposium on Antennas and Propagation and USNC/URSI National Radio Science Meeting*, San Diego, CA, USA, July 9–14, 2017, pp. 2143–2144
19. https://www.cst.de/Content/Company/Academic.aspx
20. T. Weiland, Time domain electromagnetic field computation with finite difference methods. Int. J. Numer. Modelling V Electron. Netw. Devices Fields **9**(4), 295–319 (1996)
21. X. Zhao, J. Geng, R. Jin, Y. Jin, X. Liu, W. Yin, Topological design of planar circularly polarized directional antenna with low profile using particle swarm optimization. Int. J. Antenna Propag. **2017** (2017), Article ID 4983724, 12 pages
22. H. Wu, J. Geng, R. Jin, J. Qiu, W. Liu, J. Chen, S. Liu, An improved comprehensive learning particle swarm optimization and its application to the semiautomatic design of antennas. IEEE Trans. Antennas Propag. **57**(10), 3018–3028 (2009)
23. J. Geng, R. Jin, X. Liang, H. Wu, S. Ye, B. Zhou, X. Tao, The study on the antenna optimization, in *PIERS*, Xi'an, China, 23–28 March 2010
24. Y. Fan, Y. Cui, R. Li, Polarization reconfigurable omnidirectional antenna using crossed dipoles, in *2015 IEEE International Symposium on Antennas and Propagation and USNC/URSI National Radio Science Meeting*, Vancouver, BC, Canada, July 19–14, 2015, pp. 2371–2372
25. X. Quan, R. Li, M.M. Tentzeris, A broadband omnidirectional circularly polarized antenna. IEEE Trans. Antennas Propag. **61**(5), 2363–2370 (2013)
26. W.J. Krzysztofik, Space diversity parameters of MIMO systems small antenna array for mobile terminal, in *2016 10th European Conference on Antennas and Propagation (EuCAP)*, Davos, Switzerland, Apr. 10–15, 2016, pp. 1–4
27. H. Aliakbari, A. Abdipour, A. Costanzo, D. Masotti, R. Mirzavand, P. Mousavi, Performance investigation of space diversity for a 28/38 GHz MIMO antenna (applicable to mm-wave mobile network), in *2016 Fourth International Conference on Millimeter-Wave and Terahertz Technologies (MMWaTT)*, Tehran, Iran, Dec. 20–22, 2016, pp. 41–44
28. J.F. Gonzalez, P. Padilla, J.F. Valenzuela-Valdes, J. Padilla, M. Sierra-Perez, An embedded lightweight folded printed quadrifilar helix antenna: UAV telemetry and remote control systems. IEEE Antennas Propag. Mag. **59**(3), 69–76 (2017)
29. J. Geng, R.W. Ziolkowski, K. Wang, X. Zhao, H. Zhou, G. Chenhu, X. Liang, R. Jin, Dual CP polarization diversity and space diversity antennas enabled by a compact t-shaped feed structure. IEEE Access, 27th June 2019, vol. 7, pp. 96284–96296

Chapter 6
A Circular Truncated Cone Slot Antenna with Circular Polarized Conical Beam

6.1 Introduction

Recently, the researchers pay more attention to antenna systems that radiate a conical beam with circular polarization (CP) character. For example, CP conical beams has been excited from the circular patch antennas with multi-feed probes in [1, 2] by means of a TM_{21} mode method [3]. A circular aperture antenna with single-port-feed [4] has attained a radiating flare angle of 28° and better azimuthal symmetry by exciting simultaneously both the TM_{01} and TE_{01} modes in a circular waveguide. Literature [5] reported a patch antenna with low profile and wideband that combined the TM_{01} and TM_{02} modes. However, all of the above antenna cannot easily be redesign if we need to radiate large angle beam from the boresight direction.

The radial line slot array (RLSA) antenna presented in [6] arranged perpendicular slots circularly and excite them in the rotational phase along the circumferential direction so as to make the null in boresight direction. It realized conical beam with different pointing angle by changing the distance between the slot pairs and the center of the radial waveguide. However, owning to its planar structure, the gain deteriorates as the beam angle gets larger from the boresight. In [7], by cutting 4 rounds of perpendicular slot pairs alone the axis of the coaxial cylinder, a high gain omnidirectional CP antenna is realized. It can be assumed as radiating 90° conical beam from boresight and by changing the distance between each round of slot pairs, the beam can cover other angles near 90°. However, due to the coupling between slot pairs and the restriction of the antenna length, the distance between each rounds of slot pairs cannot be too short, nor too long. So the angle coverage of the conical beam is limited. Consequently, the beam angle coverage is too.

In this chapter, taking advantage of the designs in [6, 7], we cut quasi-perpendicular rather than perpendicular slots on the circular truncated cone to produce CP conical beam owning to the gradient shape of the outer conductor. It is like in the middle state between the antenna presented in [6, 7]. Through simulation, we find that the direction of the conical beam mainly depends on the tilting angle of the out conductor as well as proper arrangement of the position of the two

J. Geng, *Omnidirectional Slots Antenna*,
https://doi.org/10.1007/978-981-15-9089-4_6

slot arrays. So we can use this form of antenna to fill up the conical beam coverage angel in [6, 7]. To achieve high gain, we use gradient coaxial line to make impedance match and equal energy distribution between two rounds of slot pairs. One typical antenna with tilting angle of 45° is optimized and fabricated to verify the idea [8, 9]. The proposed antenna has a simple mechanical structure and does not need complex feed network. Besides, as a large part of the bottom surface conductor ① (shown in Fig. 6.1b) can be removed and the inner coaxial cable can be cut short, we can move the SMA port upward, so the antenna can be mounted on the top of high speed carrier as a conformal antenna.

6.2 Antenna Design and Analysis

The CP conical beam antenna structure with slots array is shown in Fig. 6.1. The substrate between the inner and outer conductors of the truncated cone is dielectric Teflon ($\varepsilon_r = 2.1$).

A number of quasi-perpendicular slot pairs are cut on the surface of the antenna's outer conductor. One of the slots in a pair is inclined α_i clockwise and the other is inclined α_i anticlockwise ($i = 1$ for the upper round of slot pairs and $i = 2$ for the lower round), so they are quasi-perpendicular to each other. The slot pairs serve as two magnetic current elements and the length of each slot (L_i) is about $\lambda_g/2$ (λ_g is the wavelength in the medium). The distance between the two slots (Z_i) is set to be around $\lambda_g/4$ along the slant line of the outer conductor so that they are excited with 90° phase difference and two magnetic current elements combine to realize the CP property [7]. Besides, because of the gradient antenna structure, it should be noted that α_i is not exactly 45° and can be a tuning parameter to improve AR performance. The electric field on the outer conductor section at 5.7 GHz is simulated and illustrated at different time instants in Fig. 6.2. It is evident that the direction of electric field rotates circularly clockwise with almost same amplitude and the antenna radiates left hand CP wave.

There are two rounds of slot pairs arranged along the Z axis. Because the radiuses of the two round plane are not equal, the slot-pair numbers are not the same. In our design, the upper round has 6 slot pairs, which is half that of the lower round, for the sake of symmetry and more uniform gain. The length of the slots of upper round (L_1) is a little different from that of the lower round (L_2), which is good for bandwidth expansion and impedance matching. The distance between the two round plane along the slant line of the outer conductor (H_2) is set to be around λ_g, in which case the two round of slot pairs are excited in the same phase, so that the radiation is perpendicular to the outer conductor. The distance between the top surface and the upper round plane (H_1) along the slant line, the distance between the bottom surface and the lower round plane (H_3) should be as short as possible for the sake of antenna miniaturization. So we choose H_1 long enough to hold the upper round of slot pairs. However, in order to reduce reflection at end of the antenna, there should be little

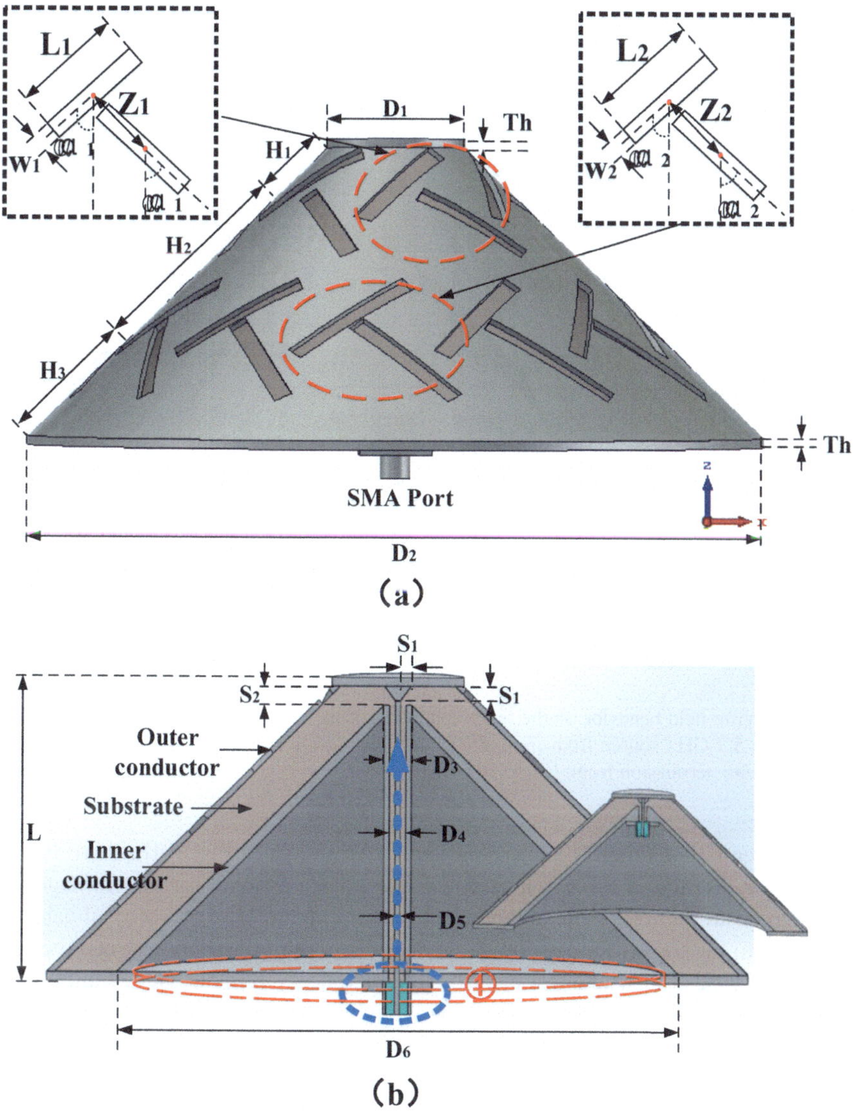

Fig. 6.1 Geometry of the CP conical beam antenna. **a** Side view. **b** Cross-sectional view. Figure reproduced with permission from: [8], © 2017 IEEE

energy left after passing through the lower round of slot pairs. Adjusting the width of the slot pairs (W_1 and W_2) and H_3 contributes to power radiation and reduces reflection.

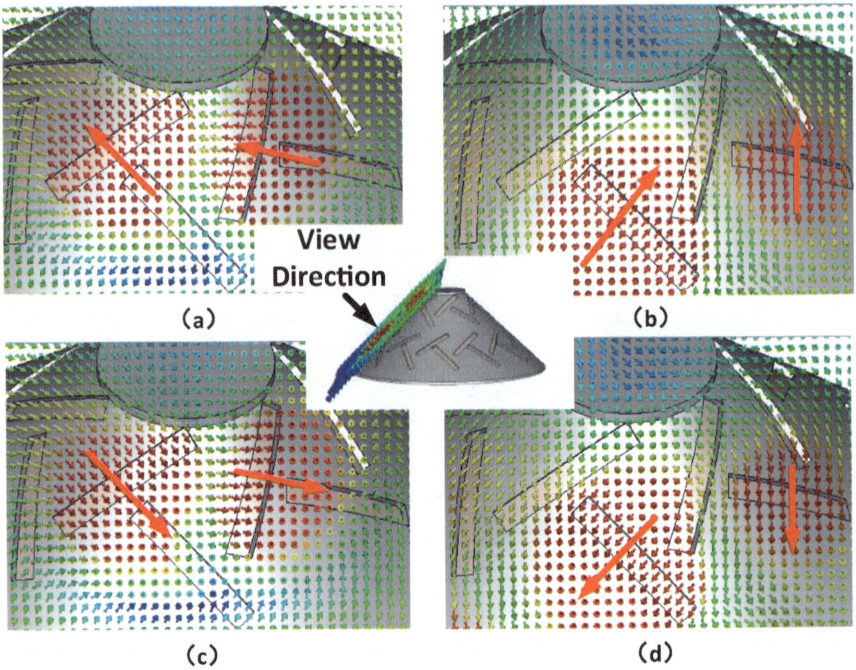

Fig. 6.2 Electric field behavior on the outer conductor of the CP conical beam antenna over one period at the 5.7 GHz source frequency. **a** $t = 0$. **b** $t = T/4$. **c** $t = 2\ T/4$. **d** $t = 3\ T/4$. Figure reproduced with permission from: [8], © 2017 IEEE

The antenna is excited by a coaxial cable. As shown in Fig. 6.1b, there is one coaxial line inserted in the antenna, which provides a channel for electromagnetic energy to go to the top of the antenna. The inner conductor of the coaxial cable is connected to the top surface of the antenna and smooth transition between them is needed, because it is where the electromagnetic energy turns around. Figure 6.3a shows the power flow of the proposed antenna operating at 5.7 GHz. After passing through the coaxial line, the energy will get through the transition area and then along the substrate to the lower part of the antenna as well as radiating from slot pairs. From cross-sectional view of Fig. 6.3b, we can notice that most of the energy is emitted from the slot arrays and there is only little amount of energy getting to the bottom of the antenna. The reflection is small so we get good traveling wave property and good AR results. It is obvious that the transition area is the key position for return loss performance. We utilize the through coaxial line to excite the slot arrays from the top of the antenna to the bottom without increasing the height of the antenna. Because if we excite the slot arrays from bottom to top, a tapered coaxial line should be added to connect the small SMA port with large bottom plane of the antenna. Apart from the transition area, the bottom diameter of the inner conductor D_6 is also a vital parameter for impendence matching. Actually, the inner conductor,

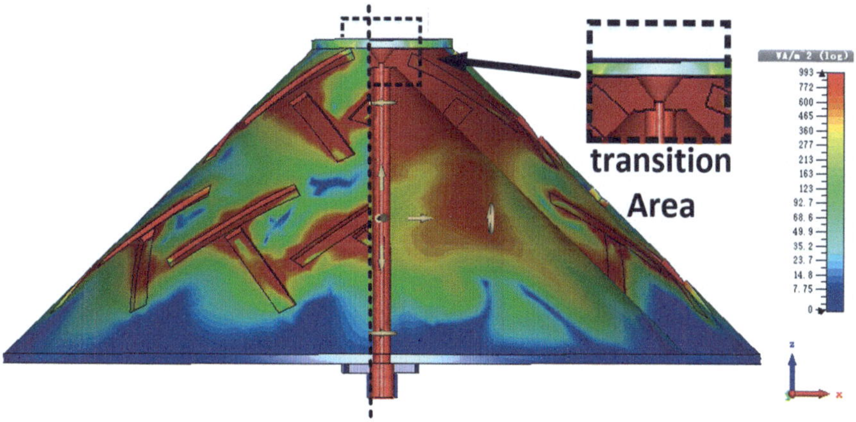

Fig. 6.3 Power flow on the outer conductor of the CP conical beam antenna at 5.7 GHz. Left: side view, and Right: cross-sectional view. Figure reproduced with permission from: [8], © 2017 IEEE

outer conductor and the substrate between them form a gradient coaxial line, which is important for equal energy radiation between two rounds of slot pairs and matching their impedance. As one end of the radiator is shorted to the ground and the other is around 50 Ω, the gradual tapered transmission path is good for adjusting the distribution impendence and keeping a relatively wide work band. The medium of the antenna is Teflon ($\varepsilon_r = 2.1$), which is filled within the area between the outer conductor of the antenna and inner conductor. For easier fabrication, we insert one existing coaxial line product into the inner conductor.

The antenna is simulated and optimized using CST Microwave Studio software based on the finite integration technique (FIT) method. The optimized geometric parameters are summarized in Table 6.1.

Table 6.1 Optimized geometric parameters of the CP conical beam antenna

Par	Value (mm)	Par	Value (mm)	Par	Value (mm)
D_1	24 mm	D_2	128 mm	D_3	4.98 mm
D_4	3 mm	D_5	0.92 mm	D_6	102 mm
Th	1.5 mm	H_1	19.8 mm	H_2	33.6 mm
H_3	15.9 mm	S_1	2.4 mm	S_2	3.2 mm
W_1	3.3 mm	L_1	22 mm	W_2	3.3 mm
L_2	24.3 mm	Z_1	14.3 mm	Z_2	14.2 mm
α_1	52.7°	α_2	50.9°	L	52 mm

Table 6.2 Performance comparison of CP conical beam antennas

Literature	AR bandwidth (%)	Profile	Max. gain (dB)	Network required	Range of the beam pointing angle
[1]	28.3	$0.084\lambda_0$	5.5	Yes	~33°
[2]	2.66	$0.069\lambda_0$	6.5	Yes	~46°
[4]	5	λ_0	7.5	Yes	~28°
[5]	14.4	$0.06\lambda_0$	4.9	No	~32°
[7]	14.5	$4.3\lambda_0$	7	No	~90°
[3]	4	$0.19\lambda_0$	3.8–6.9	Yes	35°–74°
[6]	–	$0.119\lambda_0$	3–9	No	10°–60°
Prototype	8.7	λ_0 (45°)	5.8	No	0°–90°

6.3 Parameter Analysis

As mentioned in above, the transition area is critical to the return loss results. In order to reduce refection, we add a small mental truncated cone at the bottom of the top surface. The $|S_{11}|$ with and without the small mental truncated cone are plotted in Fig. 6.4. It is obvious that by smoothing the transition area, the $|S_{11}|$ performance is

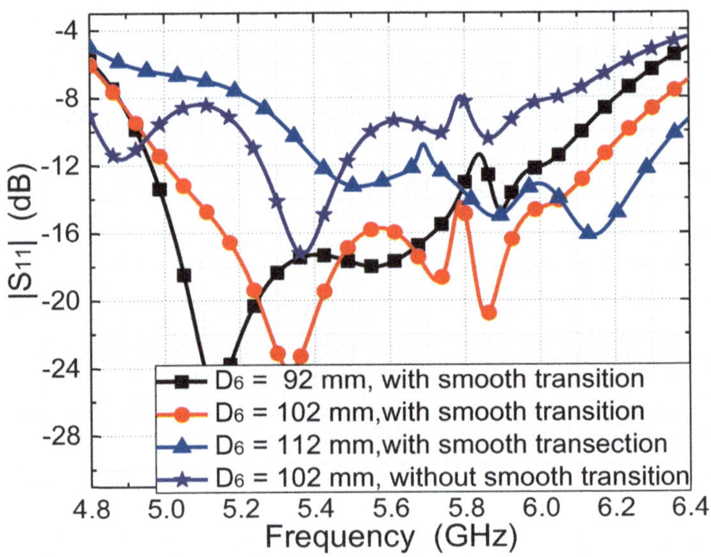

Fig. 6.4 Comparison of the simulated $|S_{11}|$ values for the CP conical beam antenna with and without the smooth transition element and the effect of bottom diameter of the inner conductor D_6 on the simulated $|S_{11}|$ values. Figure reproduced with permission from: [8], © 2017 IEEE

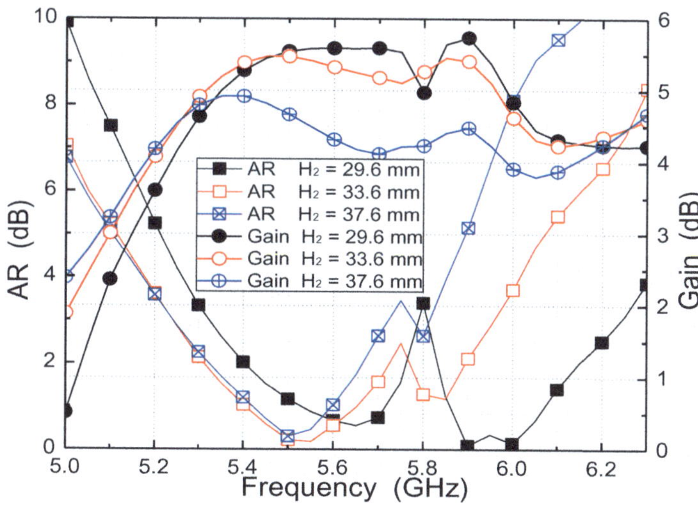

Fig. 6.5 Effect of the distance, H_2, between the two sets of slot pairs along the slant height on the simulated AR and realized gain values [8]

significantly improved. The bottom diameter of the inner conductor D_6 is also a key parameter to the impedance matching. As is shown in Fig. 6.5, $|S_{11}|$ will deteriorate if D_6 is deviated from 102 mm because of its gradient impedance characteristic. If D_6 is less than 102 mm, the $|S_{11}|$ in higher band will deteriorate, which means that the upper round of slot pairs are not properly matched. While D_6 is larger than 102 mm, the $|S_{11}|$ in lower band will deteriorate, which means that the lower round of slot pairs is not properly matched.

The gap between the perpendicular slot pairs of the upper round Z_1 and lower round Z_2 is critical to the AR performance of the antenna. With appropriate Z_1 and Z_2, the perpendicular slots are excited with 90° phase difference, so as to achieve good CP radiation. Figure 6.6 shows the effect of varied Z_1 on AR results. Any offset from the optimized Z_1 value will cause poor AR performance. The gain is also an essential parameter to antenna performance. The upper round of slot pairs and the lower round actually form a binary antenna array. So we should excite the two elements with same phase and equal power to achieve the maximum gain in the direction perpendicular to the radiator's surface. H_2 affect the phase difference between the two elements. From Fig. 6.7, we can find that H_2 not only affect the gain performance, but also AR property because it changes the current distribution. $H2 = 33.6$ mm is a compromise between the two results.

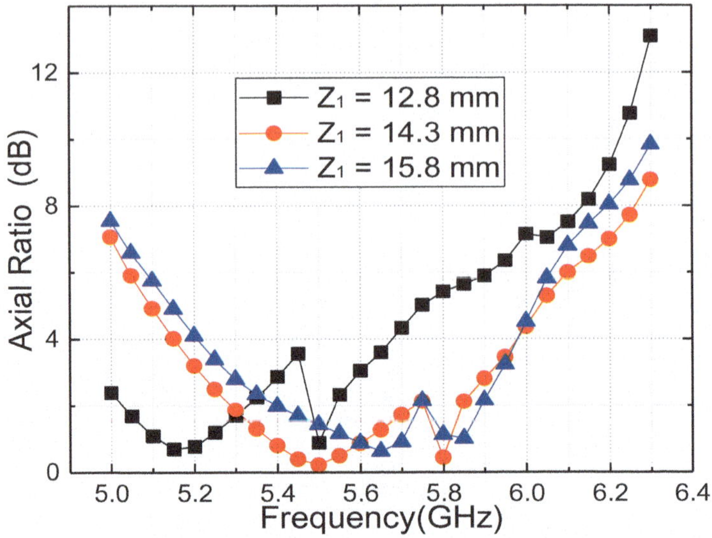

Fig. 6.6 Effect of the length between two slots of the upper set of slot pairs, Z_1, on AR results [8]

Fig. 6.7 Manufactured antenna prototype. Figure reproduced with permission from: [8], © 2017 IEEE

6.4 Results and Discussion

Simulated and measured results of the proposed antenna are presented in this section. Based on the optimized results, the prototype circular truncated cone slot antenna was fabricated. The test scenario is shown in Fig. 6.8.

It can be observed in Fig. 6.8 that the measured $|S_{11}| < -10$ dB ranges from 5.05 to 5.77 GHz, which is somewhat deteriorated compared with simulated results because of the fabrication inaccuracy and imprecise medium permittivity value. Figure 6.9 illustrates that the AR of theta = 45° is lower than 3 dB from 5.5 to 6 GHz, which is deviated to higher frequency. Because the AR is sensitive to the position of the slot pairs, it mainly contributed to the mechanical error. Within the same band, the measured left hand CP gain of antenna shows the same trend as the simulated one and it is between 4.84 and 5.8 dBic. The Measured LHCP normalized radiation pattern, the Simulated normalized LHCP and RHCP radiation pattern in the E plane at 5.5, 5.7, 5.9 GHz is shown in Fig. 6.10. From the simulated results, it is shown that the cross-polarization is about 16 dB lower than the co-polarization and the maximum gain direction remains almost unchanged and points to 45° from boresight within the band. The measured maximum gain direction is also around 45°, while in the direction of 15° from boresight, the gain is higher than the simulated one mainly because of the fabrication error and test error. The performance comparison of CP conical beam antennas are summarized in Table 6.2.

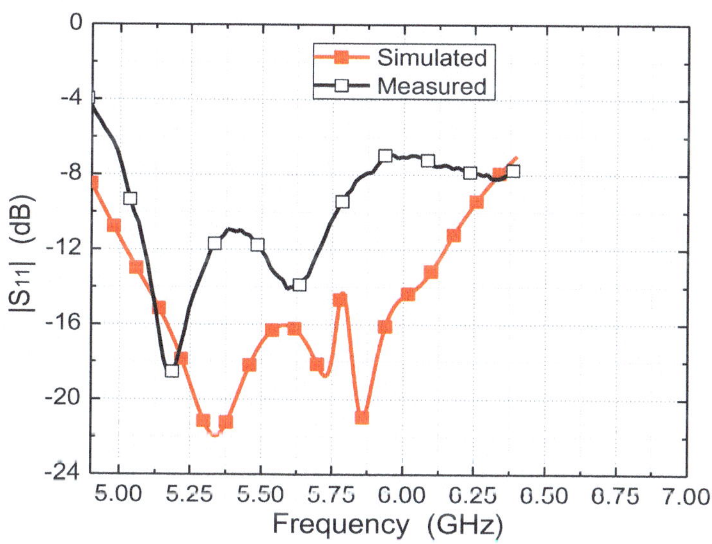

Fig. 6.8 Comparison of the simulated and measured $|S_{11}|$ values of the CP conical beam antenna. Figure reproduced with permission from: [8], © 2017 IEEE

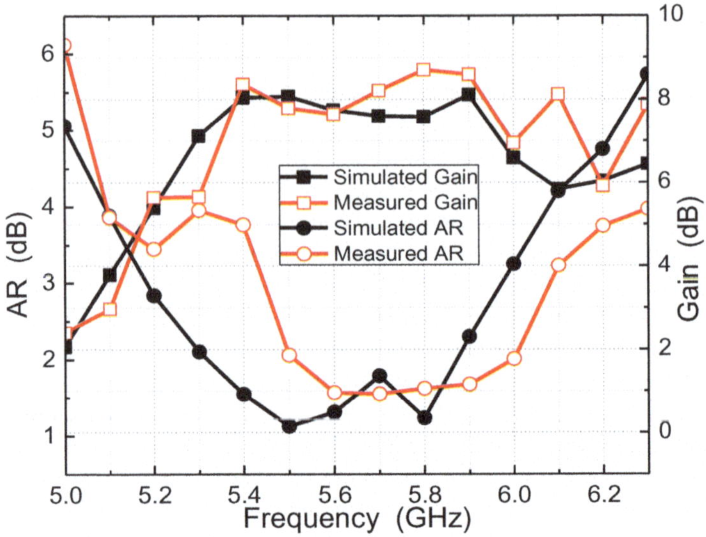

Fig. 6.9 Comparison of the simulated and measured realized gain and AR values of the CP conical beam antenna in the direction perpendicular to the outer conductor. Figure reproduced with permission from: [8], © 2017 IEEE

6.5 Summary

In this chapter, a circular truncated cone slot antenna that radiates a LHCP beam at 45° with respect to boresight was studied. The measured prototype demonstrated a 720 MHz impedance bandwidth, covering 5.05–5.77 GHz and a 500 MHz 3-dB AR bandwidth, from 5.5 to 6.0 GHz. The measured realized gain over this AR bandwidth ranged from 4.84 to 5.8 dBi. It is noted that by making one slot in each pair be inclined α_i degrees anticlockwise while making the other α_i degrees clockwise, RHCP performance can be realized, Moreover, by adding more sets of slot pairs and adjusting the width of slots, higher realized gain can be realized. Finally, the direction of the beam was demonstrated to be perpendicular to the outer conductor. Consequently, it is easy to change the beam direction simply by changing the slant angle of the outer conductor to meet the actual requirements.

A circular truncated cone slot antenna radiating 45° LHCP beam is proposed. The manufactured prototype achieved a 720 MHz impedance bandwidth, over 5.05–5.77 GHz and a 500 MHz 3-dB AR bandwidth during 5.5–6.0 GHz. The measured realized gain over the CP bandwidth is 4.84–5.8 dBi. It should be noticed that by making the slot pairs incline α_i degrees anticlockwise and α_i degrees clockwise, we can get the right hand CP performance and by adding more rounds of slot pairs and adjusting the width of slots, higher gain can be realized. Moreover, the direction of the beam is perpendicular to the outer conductor, so it is easy to change the direction of the beam by changing the tilting angle of the outer conductor to meet the actual requirements.

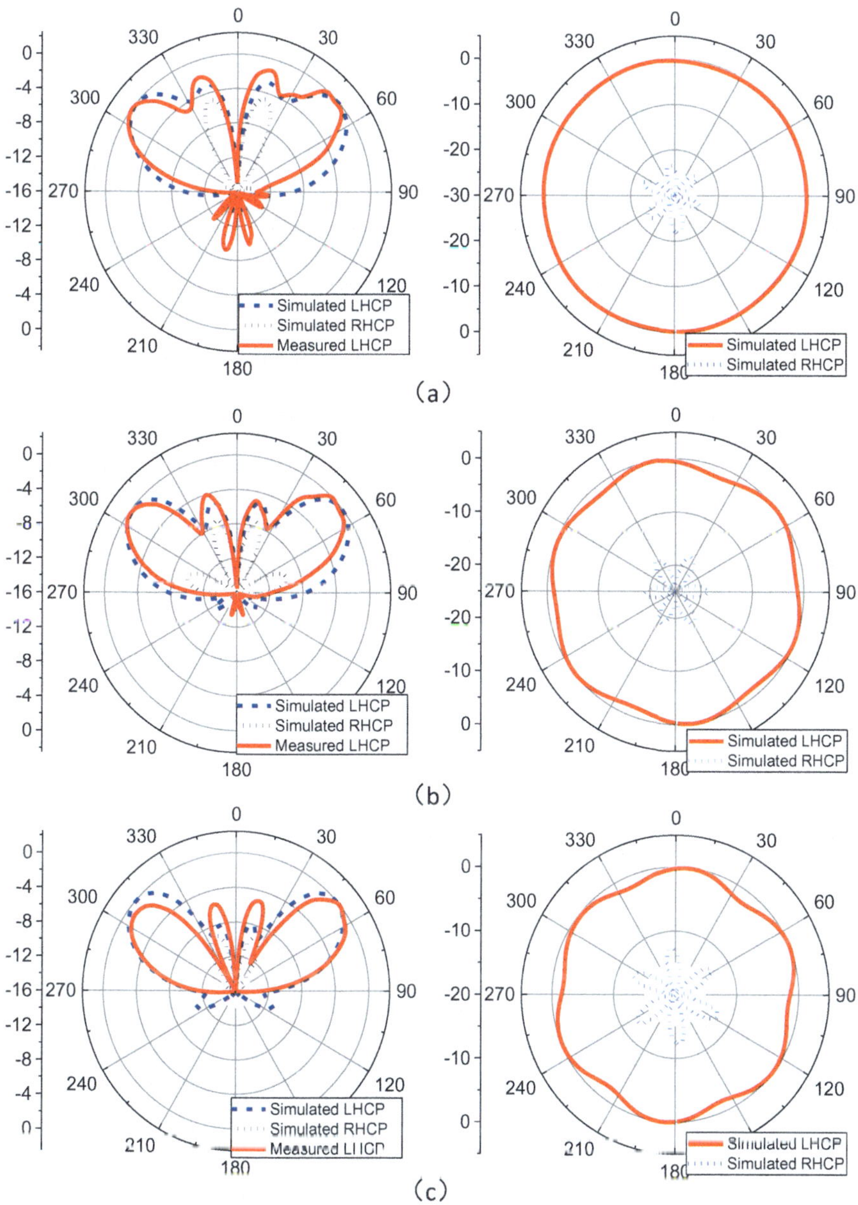

Fig. 6.10 Comparisons of the normalized, measured LHCP and simulated LHCP and RHCP realized gain patterns in the E-plane and 45° elevation angle plane at **a** 5.6 GHz, **b** 5.8 GHz, **c** 6.0 GHz. Figure reproduced with permission from: [8], © 2017 IEEE

References

1. K.L. Lau, K.M. Luk, A wideband circularly polarized conical-beam patch antenna. IEEE Trans. Antennas Propag. **54**, 1591–1594 (2006)
2. X. Bai, X. Liang, M. Li et al., Dual-circularly polarized conical-beam microstrip antenna. IEEE Antennas Wirel. Propag. Lett. **14**, 482–485 (2015)
3. J. Huang, Circularly polarized conical patterns from circular microstrip antennas. IEEE Trans. Antennas Propag. **32**, 991–994 (1984)
4. S.S. Qi, W. Wu, D.G. Fang, Singly-fed circularly polarized circular aperture antenna with conical beam. IEEE Trans. Antennas Propag. **61**, 3345–3349 (2013)
5. W. Lin, H. Wong, Circularly polarized conical-beam antenna with wide bandwidth and low profile. IEEE Trans. Antennas Propag. **62**, 5974–5982 (2014)
6. J. Takada, A. Tanisho, K. Ito et al., Circularly polarised conical beam radial line slot antenna. Electronics Lett. **30**, 1729–1730 (1994)
7. B. Zhou, J. Geng, X. Bai et al., An omnidirectional circularly polarized slot array antenna with high gain in a wide bandwidth. IEEE Antennas Wirel. Propag. Lett. **14**, 666–669 (2015)
8. G. Chenhu, J. Geng, H. Zhou, J. Li, L. liu, X. Liang, W. Zhu, R. Jin, R.W. Ziolkowski, Truncated circular cone, slot antenna array that radiates a circular polarized conical beam. IEEE Antennas Wirel. Propag. Lett. **16**, 2574–2577 (2017)
9. G. Chenhu, J. Geng, L. Liu, H. Zhou, X. Zhao, Y. Liang, X. Liang, W. Zhu, R. Jin, A circular truncated cone slot antenna with circular polarized conical beam, in *2017 IEEE International Symposium on Antennas and Propagation and USNC/URSI National Radio Science Meeting*, San Diego, CA, USA, July 9–14, 2017, pp. 1533–1534

Chapter 7
Half-Space Covered Antenna for Air-Ground Communication

7.1 Introduction

In recent years, people have become more and more interested in the field of air-ground communications, such as communication between ground command stations and drones. Under normal circumstances, air-ground communication systems require antennas with wide coverage beams and circular polarization performance. It is very necessary to consider the attenuation of electromagnetic waves in different elevation planes into the antenna design. On the one hand, increasing the horizontal distance will make the elevation angle smaller, and at the same time make the air-ground passage more complicated. On the other hand, the increase in the distance between the transmitter and the receiver will increase the loss in free space [1, 2]. In summary, the attenuation of electromagnetic waves in the horizontal direction is much greater than in other directions. In order to compensate for the attenuation difference at different elevation angles, the half-space coverage antenna should have a higher gain in the horizontal plane. As shown in Fig. 7.1, the antenna proposed in [3] can achieve excellent axial ratio and relatively high gain in the horizontal direction. The disadvantage is that its axial ratio in the boresight direction is zero, and the radiation pattern is unstable. The embedded folded four-quardrifilar helix antenna proposed in [4] can obtain a wide range of elevation angles while ensuring good axial ratio performance. But its radiation pattern is hemispherical, which makes the main lobe point in the direction of the boresight. The gain in the horizontal direction is small.

Based on the advantages of [3], this chapter proposes a slot array antenna with a spiral structure on the top [5]. The proposed antenna compensates for the top null of the radiation pattern by pulling out the inner conductor and making it spiral. It can obtain excellent AR results and high gain in the horizontal direction. Therefore, the half-space covered antenna is very suitable for air-ground communication.

J. Geng, *Omnidirectional Slots Antenna*,
https://doi.org/10.1007/978-981-15-9089-4_7

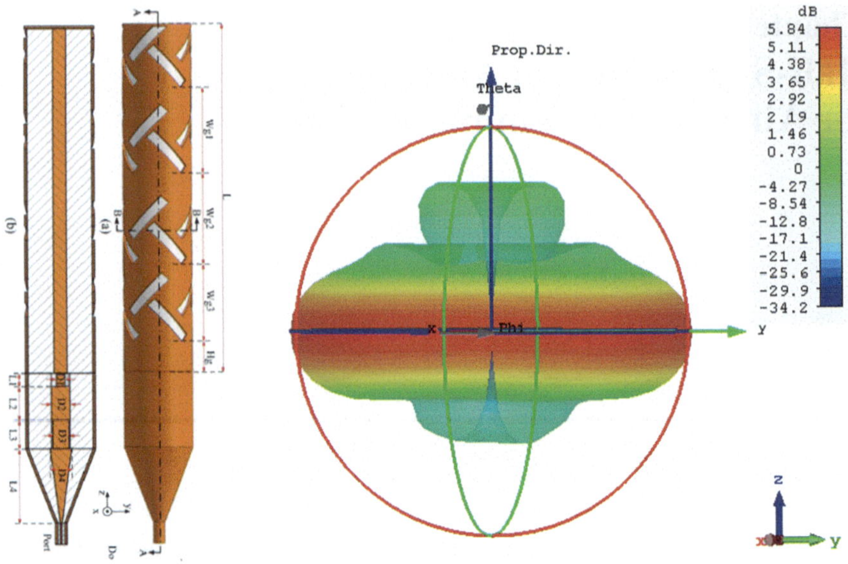

(a) (b)

Fig. 7.1 Omnidirectional CP antenna, **a** OCPA structure and **b** far field

7.1.1 The Principle of the Half Space Covering Antenna

As shown in Fig. 7.2 [5], the proposed antenna can be feed by SMA port directly. On the outer conductor of the coaxial line, two sets of slot pairs are cut in order to achieve high gain in the horizontal direction. Two pair of slots are the same in size, with length about $\lambda_g/2$ (λ_g is the wavelength in the substrate between the inner and outer conductor). The dielectric between the inner conductor and outer radiator is chosen as Teflon ($\varepsilon_r = 2.1$). A gradual tapered coaxial line is taken as a transition between the SMA port and the coaxial line to make up for the difference of size and achieve excellent impedance matching. Firstly, the two slots in one pair are perpendicular to each other in position. Secondly, the distance between the center points of the two slots is about $\lambda_g/4$, which make them excited with a phase difference of 90°. Therefore, the two slots in one pair constitute the basic unit of the CP. The two sets of slot pairs combined to form an array, the length between them is set to be about λ_g so that they are excited with same phase and achieve high gain in horizontal direction. The gain fluctuation above the horizontal plane needs to be compensated by the middle radiation part. The remaining energy of the bottom radiating part also needs the middle part to act as a load. The number of slit pairs in each group on a circular plane is nine. Due to space constraints, the length of the pair of slits on the middle radiating part is smaller than the length of the pair of slits on the bottom radiating part.

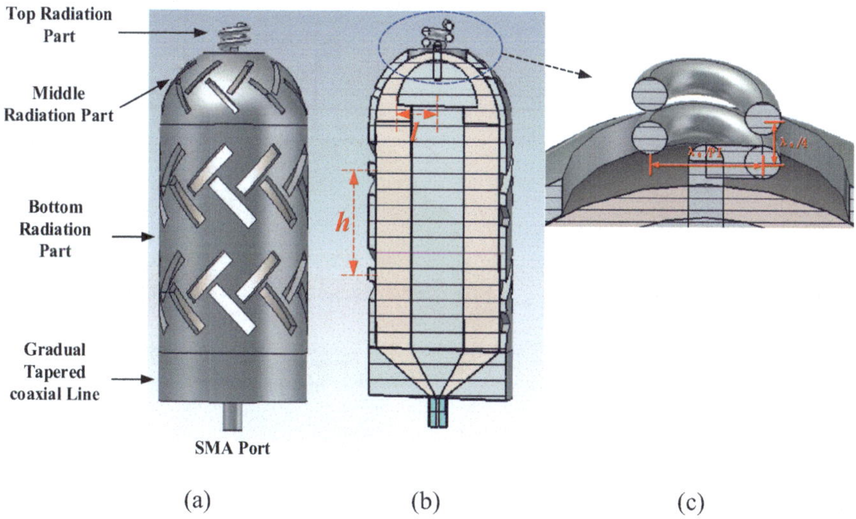

Fig. 7.2 Geometry structure of the half-space covered antenna. **a** Side view, **b** cross-sectional view, and **c** partial enlarged figure of the top helix [5]

In Fig. 7.3, from side direction, the electric field distribution in the 2D plane mainly rotates along right-hand circular direction in Fig. 7.3a–d with time $t = 0$, T/4, 2 T/4, 3 T/4 in turn. So that the radiated wave in horizontal plane is omnidirectional RHCP wave. As shown in Fig. 7.3a, the power flow distribution again confirms that the energy radiated along the radial direction in the horizontal plane.

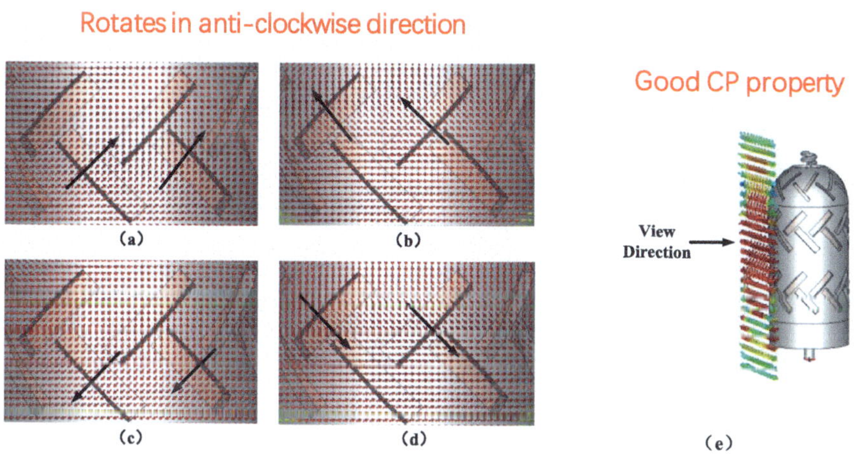

Fig. 7.3 Electric field behavior on the outer conductor of the half-space covering antenna over one period at the f_0 GHz source frequency. **a** $t = 0$, **b** $t =$ T/4, **c** $t = 2$ T/4, **d** $t = 3$ T/4

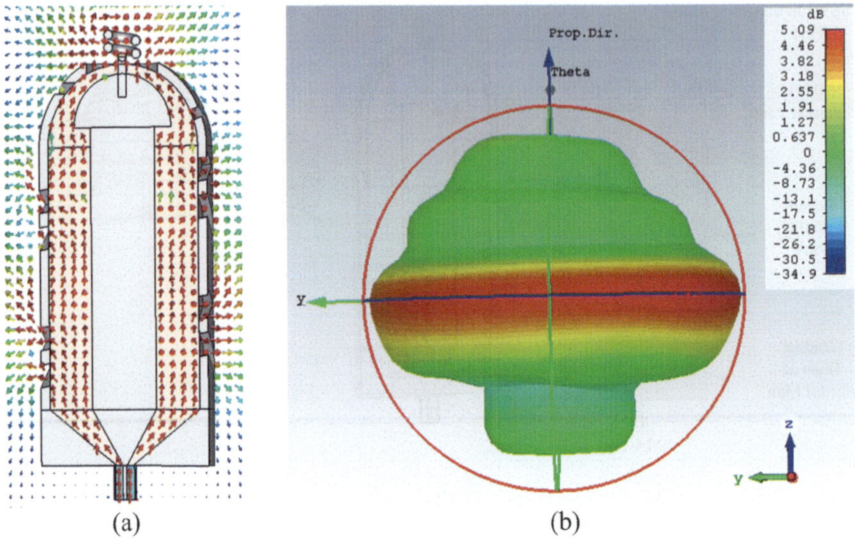

Fig. 7.4 Radiation performances, **a** power flow, **b** half-space covering pattern [5]

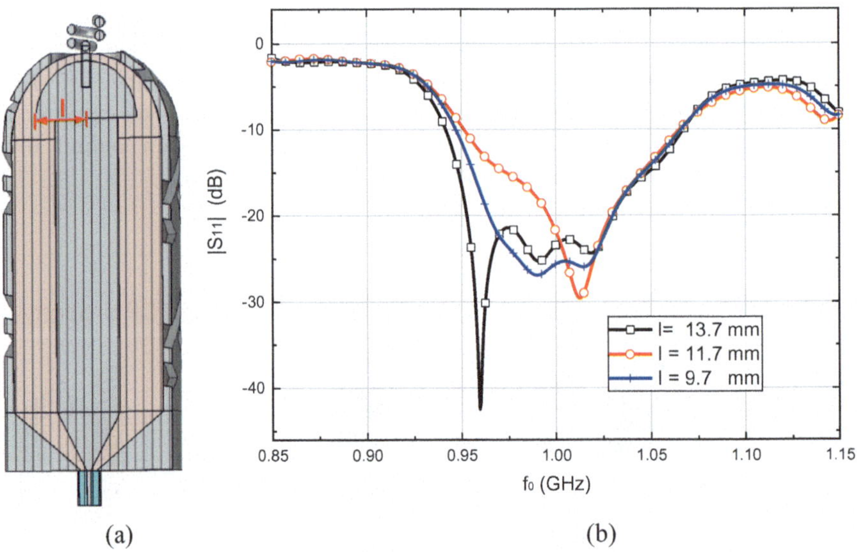

Fig. 7.5 Effect of varied radius of the top inner conductor, *l*, on $|S_{11}|$. **a** Parameter, **b** $|S_{11}|$

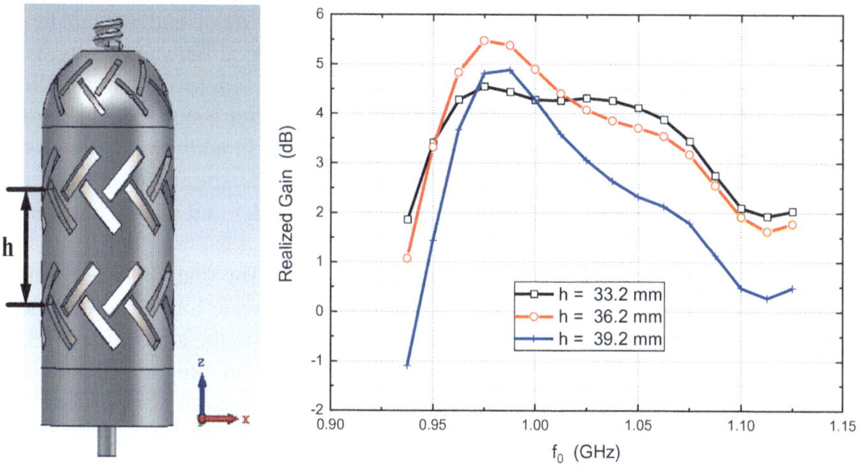

(a) (b)

Fig. 7.6 Effect of varied length *h* between the two sets of slot pairs on the bottom radiation part, **a** diagram of the antenna, **b** *h*, on realized gain results

Fig. 7.7 The simulated $|S_{11}|$ results of the proposed antenna

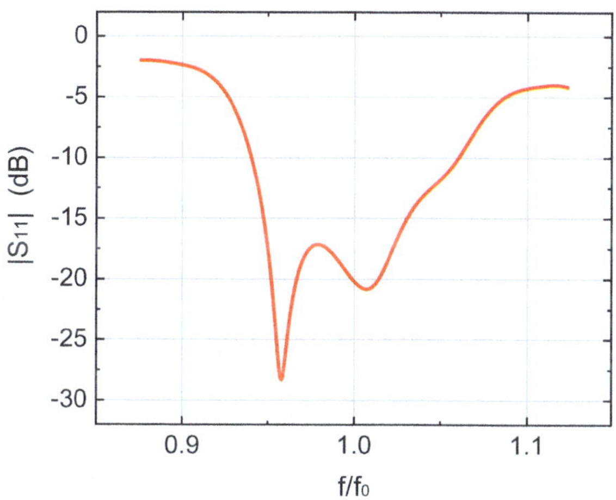

Since each pair of slits is excited by the rotational phase, the boresight direction of the radiation pattern will be zero. However, there is little energy at the top of the outer conductor, and there is not enough space at the top to cut slots. To solve this problem, a hole is reserved at the top of the outer conductor, the purpose of which is to pull out the inner conductor of the antenna and make the inner conductor spiral. As shown in Fig. 7.3, the end of the inner conductor is a helix, which extend out to the top of the antenna through a larger hole on the out shell of the cylinder end. The circumference of one turn is set to be around λ_0 (λ_0 is the wavelength in the air), and the diameter of the helix is about. The helix runs in an axial mode and its main beam is directed towards the axis of the helix. The vertical interval between the turns is set to approximately $\lambda_0/4$, and the number of turns is 2. In order to obtain excellent impedance matching performance, a hemispherical conductor is connected between the inner conductor of the coaxial line and the spiral line. In addition, although the energy density in the transition zone is not as great as the spirals or slots on the coaxial cylinder, a few slot pairs near the top of the cylinder end also radiate some waves.

Then the whole far field pattern is shown in Fig. 7.3b. The main lobe is omnidirectional in the horizontal plane, and the upper lobe has a wide lobe without a clear null point, so it is like a half space, covering the null point in the z-axis direction.

Usually, the z axis direction mainly points to upward the moving terminal in sky, and the distance of the moving terminal just on the right above is nearest to the ground. So the gain in this direction is smallest to satisfy the channel gain, even a null point is enough too.

7.1.2 The Key Parameters of the Half Space Covering Antenna

After simulation, it found that the radius l of the top half sphere of the inner conductor and the interval h between two round slots mainly affect the antenna performances.

(1) The radius l of the top half sphere of the inner conductor

As shown in Fig. 7.4a, this antenna is a serial radiating antenna with the slots rounds and the helix in turn. And the top half sphere is just to control the power distribution to the beginning slots and the end helix firstly, so that the shape of the pattern will change with the parameter l.

In another side, for the adjustment function to radiation, the return loss can be changed by l too. As shown in Fig. 7.5, the available work band have little difference for l varies from 9.7 to 13.7 mm, but the pit on the S_{11} curve appear at lower frequency at $l = 13.7$ mm, move to higher frequency when $l = 11.7$ mm, and the pit become shallow when $l = 9.7$ mm.

(2) The interval length between two slots rounds

As shown in Fig. 7.6, these two slots rounds is the main radiation part of the antenna, and the beam is mainly in horizontal plane. As discussed in Chap. 3, the distance h decide the inner wave guide distance and the phase difference between these two slots rounds. In another side, if we hope the combined beam from these two slots rounds distributed in the *xoy* plane (horizontal plane), the phase difference between these two slots rounds should be smaller than 90°.

Here the realized gains (in horizontal plane) between different values of parameter h are shown in Fig. 7.6b. With the parameter h increases from 33.2 to 39.2 mm, the peak gain increases firstly, and then decreases. And the gain peak is about 5.5 dBi at $0.975f_0$ GHz when $h = 36.2$ mm. And the frequency band with high gain decreases when the parameter h increases.

7.2 Performance Analysis

The proposed antenna is simulated and optimized with CST Microwave Studio software. The simulated results are presented in this section. As shown in Fig. 7.7, the −10 dB impedance bandwidth ($|S_{11}| < -10$ dB) of the proposed antenna is about 12.5%, covering from $0.9375f_0$ to $1.0625f_0$ GHz (f_0 is the center operation frequency of the proposed antenna). The 3 dB AR bandwidth in horizontal plane is 7.5%, covering from $0.9625f_0$ to $1.0375f_0$, which is shown in Fig. 7.8. And the realized gain is higher than 3.5 dB from $0.9625f_0$ to $1.0625f_0$ GHz. The radiation pattern of the half-space covered antenna at$0.95f_0$, f_0, and $1.05f_0$ are shown in Fig. 7.9. It is clear that the realized gain of the proposed antenna in the elevation angle from 90°to 0°changes mildly, so the beam covering property is good. The realized gain in boresight direction is higher than −16 dB during $0.9125f_0$ to f_0. The normalized right hand CP (RHCP) radiation pattern and left hand CP (LHCP) in E plane at f_0 GHz is shown in Fig. 7.10. The cross-polarization in the horizontal plane is at least 16 dB lower than co-polarization, so it achieves good AR results in horizontal direction.

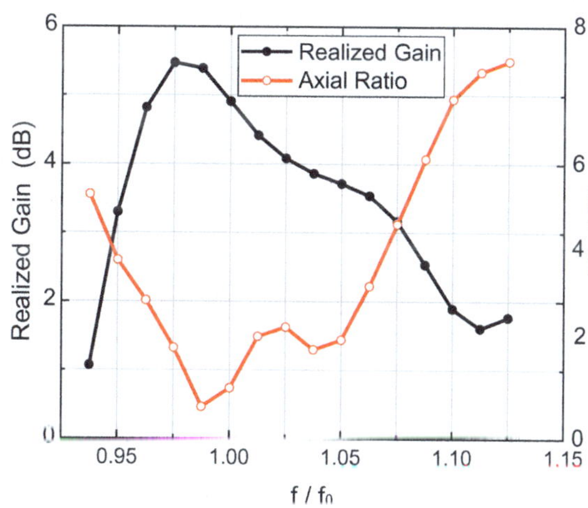

Fig. 7.8 The simulated realized gain and AR results

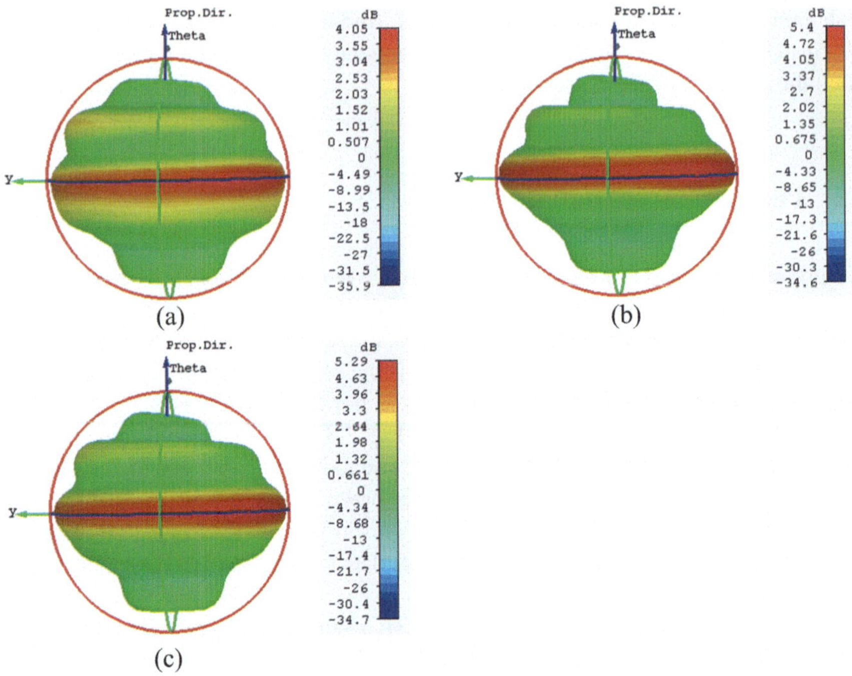

Fig. 7.9 The radiation pattern of the half space covering antenna at **a** $0.95f_0$, **b** f_0, **c** $1.05f_0$

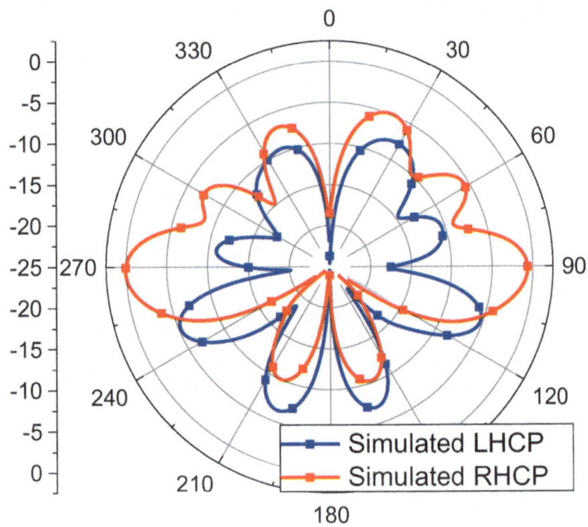

Fig. 7.10 The normalized RHCP and LHCP radiation pattern in the E plane of the half-space covered antenna at f_0

7.3 Summary

A half-space covered antenna with horizontal right hand circular polarization (RHCP) property is presented, which is realized by perpendicular slot pairs cut on the outer conductors of one coaxial line. The proposed antenna achieves high RHCP gain and good AR on the horizontal plane. As a result, it can compensate for the attenuation of electromagnetic waves. It has good beam coverage characteristics, which manifests itself as small changes in RHCP gain on a higher elevation plane. In order to cover the null point of the radiation pattern in the direction of the viewing axis, the inner conductor of the antenna is drawn out and made into a spiral shape. As a transition between the slot array and the spiral structure, the hemispherical outer conductor is also cut by the slot pair. It is worth noting that for air-ground communication, the attenuation of electromagnetic waves in the horizontal plane is much greater than in other directions, and the greater the deviation from the plane, the smaller the attenuation.

The proposed antenna performs a -10 dB impedance bandwidth ($|S_{11}| < -10$ dB) of 12.5% from $0.9375f_0$ to $1.0625f_0$ GHz and 3 dB AR bandwidth of 7.5% from $0.9625f_0$ to $1.0625f_0$ GHz. It should be noted that the gains in the horizontal direction and the boresight direction need to be compromised. On the one hand, in order to obtain higher gain on the horizontal plane, it is inevitable to cut more slit pairs in the bottom radiation part. But this will result in attenuation of the gain in the boresight direction. On the other hand, the spiral at the top of the antenna will deteriorate the horizontal AR results. Therefore, it is very important to design these two parts reasonably according to actual application requirements.

References

1. G. Maral, M. Bousquet, *Satellite Communication Systems* (Willey, Chichester, England, 2002)
2. M. Richharia, *Satellite Communication Systems* (McGraw Hill, New York, 1999)
3. B. Zhou, J. Geng, X. Bai et al., An omnidirectional circularly polarized slot array antenna with high gain in a wide bandwidth. IEEE Antennas Wirel. Propag. Lett. **14**, 666–669 (2015)
4. J.M.F Gonzalez, P. Padilla, J. Valenzuela-Valdes, et al., An embedded lightweight folded printed quadrifilar Helix antenna: UAV telemetry and remote control systems. IEEE Antennas Propag. Maga. 1–8 (2017)
5. G. Chenhu, J. Geng, H. Zhou, J. Li, L. Liu, Y. Chen, Y. Liang, M. Zhu, X. Liang, W. Zhu, R. Jin, A half-space covered antenna for air-ground communication, in *2017 Sixth Asia-Pacific Conference on Antennas and Propagation (APCAP)*, 16–19 Oct. 2017, Xi'an, China

Chapter 8
A Compact Reconfigurable Coaxial Slot Antenna

8.1 Introduction

Reconfigurable antenna, due to their ability to obtain directive radiation pattern have been widely used in wireless communication system. There are various methods for beam-scanning antenna in the literature [1–4]. Most of them cannot steer beam to cover the entire horizontal plan [1, 2]. In the literature [3] and [4], the proposed antenna have the ability of steering beam to sweep in the entire horizontal plane but the antenna need many active element, which lead to more costs and more complexity to fabrication.

In this chapter, a compact reconfigurable coaxial slot antenna based on coaxial cylinder structure with slot array and switch structure is proposed, which can realize directive radiation pattern that cover the entire horizontal plane by controlling the state of PIN diodes which is lapped at the midpoint of coaxial slot.

8.2 Design Principle

The main mode of coaxial line is TEM, which base the dual transmission system. Figure 8.1 is a geometry of the coaxial line. The diameter of inner conductor is a, and outer conductor is *b*. The electromagnetic field on the cross-section can be solved using a two-dimensional static equation. The expression of the electromagnetic field transmitted on the same axis is:

$$\vec{E} = \hat{a}_r \frac{V_r}{r \ln\left(\frac{b}{a}\right)} e^{-jkz} \tag{8.1}$$

$$\vec{H} = \hat{a}_\varphi \frac{V_r}{r \ln\left(\frac{b}{a}\right) Z_{TEM}} e^{-jkz} \tag{8.2}$$

J. Geng, *Omnidirectional Slots Antenna*,
https://doi.org/10.1007/978-981-15-9089-4_8

Fig. 8.1 The coaxial line in cylindrical coordinate

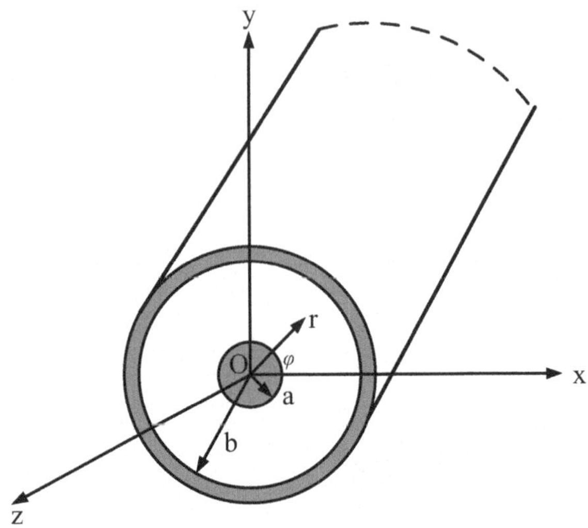

The surface current of the outer conductor is:

$$\vec{J} = -\hat{a}_r \times \vec{H} = \hat{a}_z \frac{V_r}{r \ln\left(\frac{b}{a}\right) Z_{TEM}} e^{-jkz} \tag{8.3}$$

It can be seen from Eq. (8.3) that the surface current on the outer conductor of the coaxial line is distributed along the z axis of the cylindrical coordinate system. Therefore, a moderately etched slot on the coaxial line can cut the surface current and excite the slot to radiate. This is the basic principle of the design in this section.

The model of the antenna is shown in Fig. 8.2. Figure 8.2a is the structure diagram of the antenna, and Fig. 8.2b is a perspective view of the antenna structure. The substrate of the antenna is Teflon, and its dielectric constant is 2.1. The basic structure of the antenna is a coaxial line. The substrate is filled between the inner and outer conductors. The upper end is short-circuited with the metal patch and the inner and outer conductors. The lower end is connected to the 50 coaxial matching line. The antenna can be roughly divided into two parts. The first part is a coaxial radiator etched with a slot. The long side of the slot effectively cuts the surface current on the outer conductor of the coaxial line, and then radiate.

As shown in Fig. 8.2a, the length of the etching slot is L, the width is w. and the length of the slot is approximate $\lambda/2$. According to the principle of duality, its far-field radiation is similar to the dipole. The angle between the slot and the z-axis is α, where $\alpha \approx 45°$. Since one of the slot is deviated clockwise and the other slot is deviated counterclockwise, so, the angle between the two slot is about 90°. We consider two slot as a group, and distribute four groups in the coaxial circumferential direction. Therefore, each group controls a range of about 60° around a circle. Looking at this loop of slot as a group, four groups are distributed along the coaxial z direction. These four groups can be regarded as an array. The amplitude of each array element

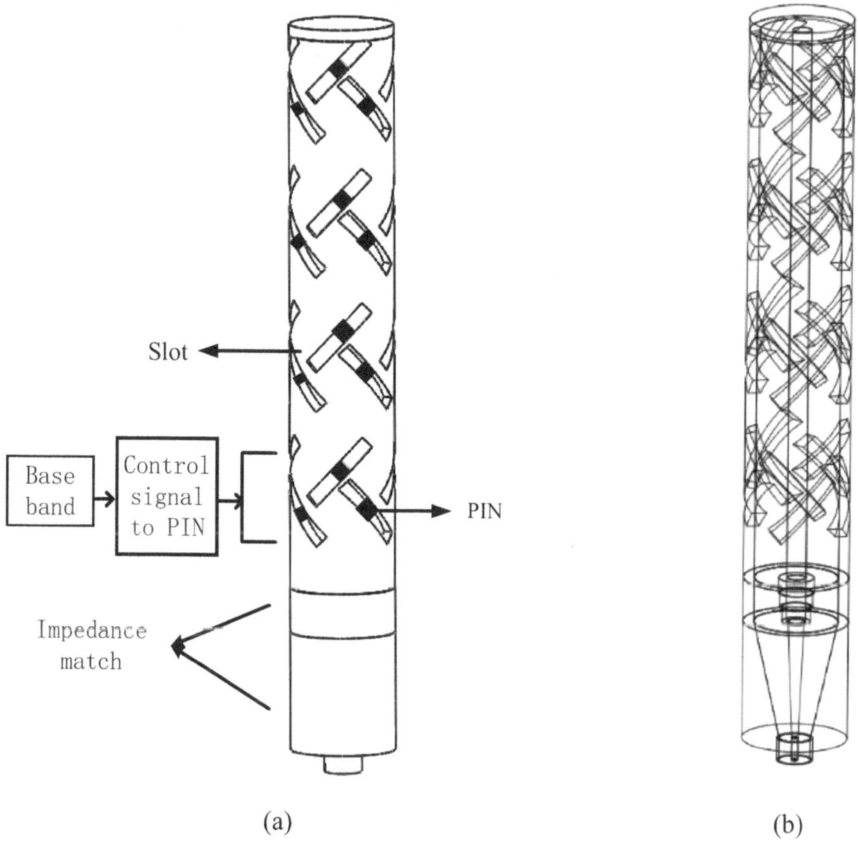

Fig. 8.2 The configuration of the proposed reconfiguration coaxial slot antenna. **a** 3-D view. **b** Phantom view. Figure reproduced with permission from: [6], © 2018 IEEE

decreases. The spacing of each group is set according to the principle of the array antenna. After optimizing, the spacing between the array elements is about 0.7λ. A PIN diode is mounted at the midpoint of the slot, and is fed by DC bias. The characteristics of the mounted diodes are changed by controlling the on and off of, and then the reconfigurable performance of the antenna is achieved. Figure 8.3 is a schematic diagram of a normal gap pair and the extreme slot pair. It can be seen that the normal slot pair has a complete rectangular slot structure, but the extreme slot is cut by short patch.

Figure 8.4 is a schematic diagram of the slot connecting different components, which shows three different gap systems. When the gap is connected with different components, the gap will show different radiation characteristic. When a small resistor is connected to the slot, since the RF current can effectively pass through the slot, the small resistor can be equivalent to a complete metal. At this time, the

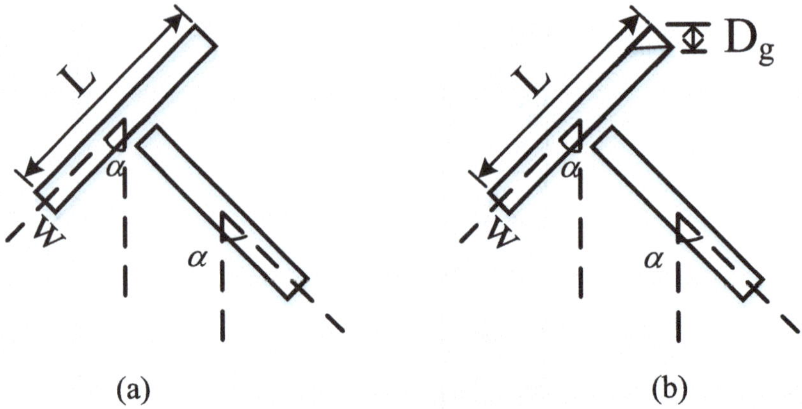

Fig. 8.3 Geometry of the slot pairs on coaxial line. **a** Normal slot pair **b** modified slot pair

Fig. 8.4 The model of slot placed different element. **a** PIN diode. **b** Resistance. **c** Capacitance

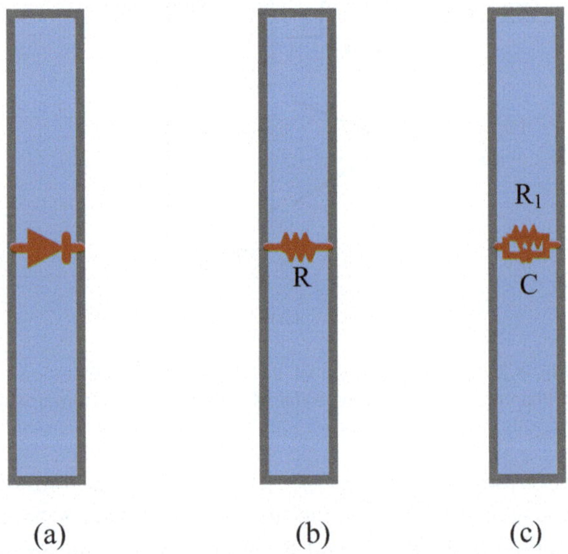

equivalent length of this slot becomes half of the original, corresponding to the resonance frequency becomes twice the target frequency and does not radiate at the target frequency. When a capacitor is connected in the slot, it can be considered that the resonance length of the slot is unchanged, and it can still radiate at the target frequency. Based on this principle, using the characteristics of the PIN diode, combined with the coaxial slot antenna, a directional pattern reconfigurable antenna can be achieved.

Figure 8.5 shows the slot with the PIN diode. According to the characteristics of the PIN diode, when the PIN diode is forward biased, it can be seen as a small resistance, and when the PIN diode is reverse biased, it can be equivalent to a parallel connection of a capacitor and resistance. According to the above analysis, the Fig. 8.5b, c show

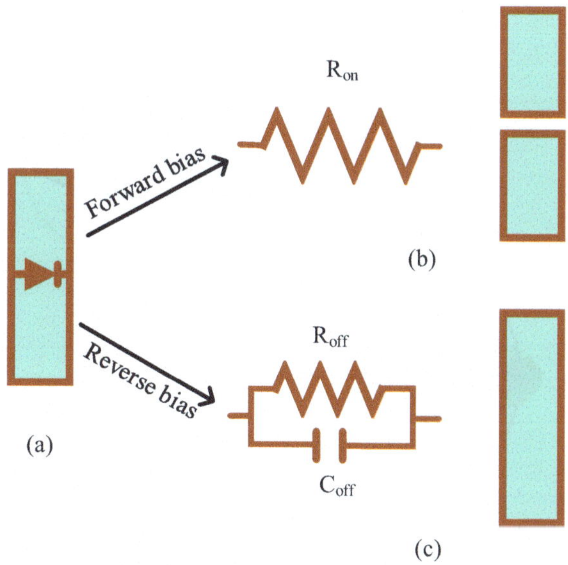

Fig. 8.5 Structure and states. **a** Geometry of the proposed slot structure. **b** On-state equivalent circuit and equivalent length of slot. **c** Off-state equivalent circuit and equivalent length of slot. Figure reproduced with permission from: [6], © 2018 IEEE

the geometry of the equivalent slots when the PIN diode is forward biased and reversed. When the PIN diode is forward biased, a complete slot will be equivalent to two slots. At this time, the slots do not radiate at operating frequency. When the PIN diode is reverse biased, the slots radiate at operating frequency.

8.3 Simulation Results

For the convenience analysis in the latter, the slot in each column are numbered 1, 2, 3, 4. The first slot 1 corresponds to a main beam direction of 0°, the second column corresponds to a main beam direction of 90°, and so on. The third slot and the fourth slot correspond to the 180° beam and the 270° beam, respectively. At this time, if the PIN diodes, which are placed on the slot of the 2nd, 3rd and 4th columns, is forward biased, and the PIN diodes of the slot on the 1st column is reversed biased, the direction corresponding to the slot of the 2nd, 3rd and 4th columns do not radiate at the target frequency and for the slot array 1, it radiates normally at the target frequency. At this time, the main radiation direction controlled by slot array 1 is 0°, and the 3 dB beam coverage is about 90°. Therefore, in the same situation, if the PIN diodes in column 2 are reversed biased, and the rest are forward biased, the beam is switched to the 90° direction. By analogy, the beam reconfiguration of 360° in the horizontal plane can be realized. Figure 8.6 show the radiation pattern of the reconfigurable coaxial slot antenna. It can be seen that the antenna can well perform omnidirectional scanning in the horizontal plane. When the forward and reverse bias of the PIN diodes are controlled, the antenna realizes the beam scanning

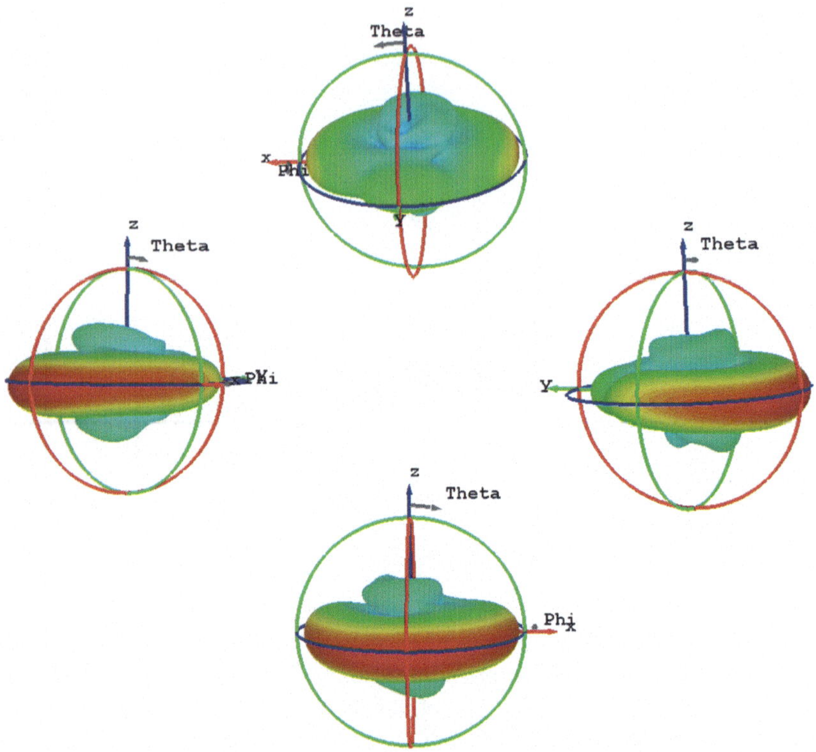

(a)

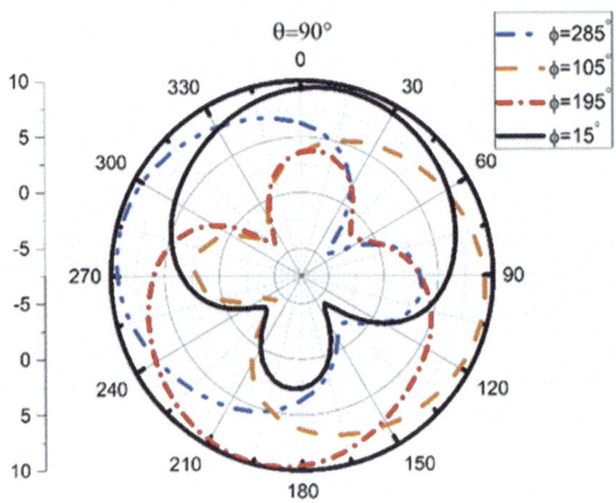

(b)

Fig. 8.6 The reconfigurable radiation pattern at 11.2 GHz. **a** 3-D view **b** Radiation pattern in azimuth plane. Figure reproduced with permission from: [6], © 2018 IEEE characteristic. And the main lobe direction are 15°, 105°, 195°, 285°. The simulated 3-D view of the reconfigurable radiation pattern at 11.1 GHz and 11.4 GHz is shown in the Fig. 8.7, and the 2-D view of radiation pattern is shown in Fig. 8.8. It can be observed that the characteristic of the reconfigurable pattern is stable in the operating frequency.

Figure 8.10 shows the simulated energy distribution. It can be seen from Fig. 8.5, the electromagnetic energy is fed from the bottom into the coaxial line, and the energy is radiated by the slot, which the PIN diodes are reverse biased. But at the PIN diodes are forward bias, the slot does not radiate, and the energy is bounded inside the coaxial line.

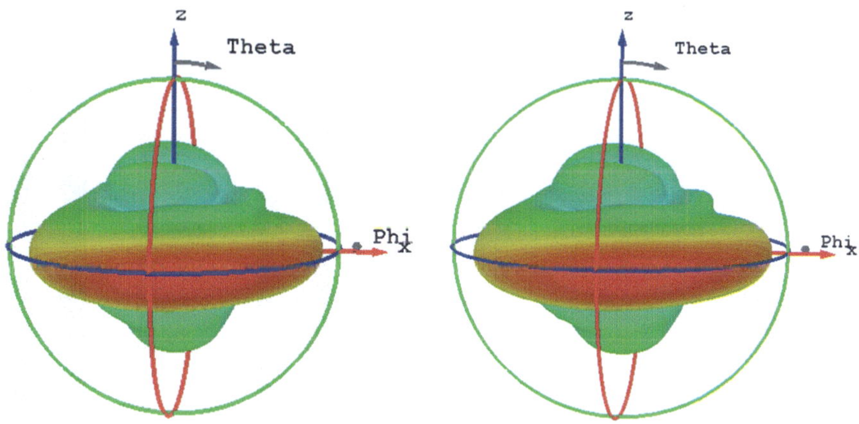

(a) (b)

Fig. 8.7 The 3-D view of the reconfigurable radiation pattern. **a** 11.1 GHz **b** 11.4 GHz

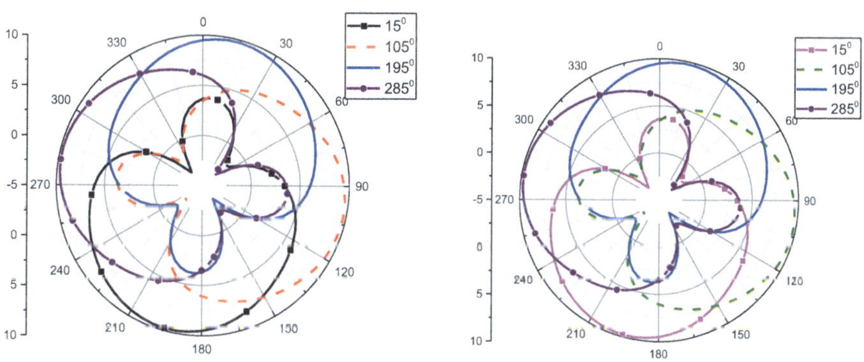

(a) (b)

Fig. 8.8 The reconfigurable radiation pattern. **a** 11.1 GHz **b** 11.4 GHz

The simulated S_{11} is shown in Fig. 8.9. It can be observed that the 10 dB impedance bandwidth is 11.08–11.45 GHz. Figure 8.11 shows gain at its main lobe direction. It can be seen from Fig. 8.11, the gain of the antenna is stable at operating the frequency, the antenna gain is basically greater than 8.7 dB, and the maximum gain is 9.75 dB. The pattern reconfigurable antenna has good high gain characteristics.

Figure 8.5 shows the structure of the slot pair etched on the coaxial line. It can be seen that the normal slot is a rectangular rectangle placed obliquely, but the last circle is cut away D_g. According to the basic principle of the transmission line, a coaxial line with a short-circuited end will form a standing wave and the electromagnetic wave is total reflected at the short end, which greatly increase cross polarization and make the reflection coefficient worse. Therefore, the purpose of setting a modified slot pair at the end is: first, adding a slot pairs will increase the overall antenna gain;

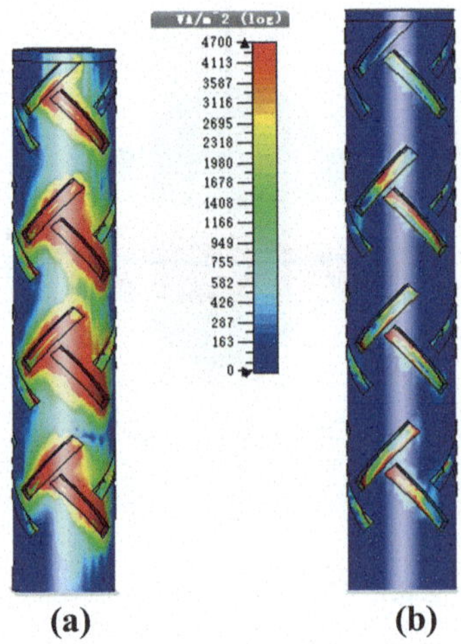

Fig. 8.9 Power flow on the out conductor of the reconfigurable pattern antenna. **a** Radiation slot **b** Not radiation slot

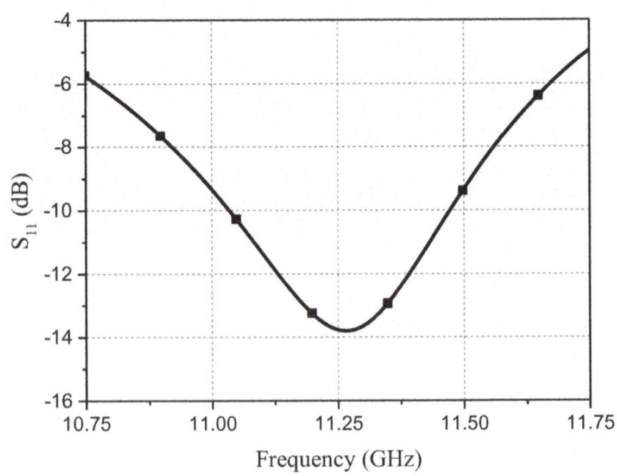

Fig. 8.10 The *S11* of the reconfigurable pattern antenna

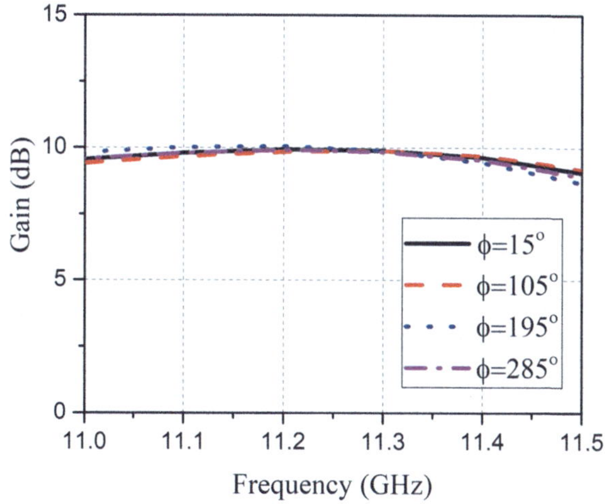

Fig. 8.11 The gain of the reconfigurable pattern antenna. Figure reproduced with permission from: [6], © 2018 IEEE.

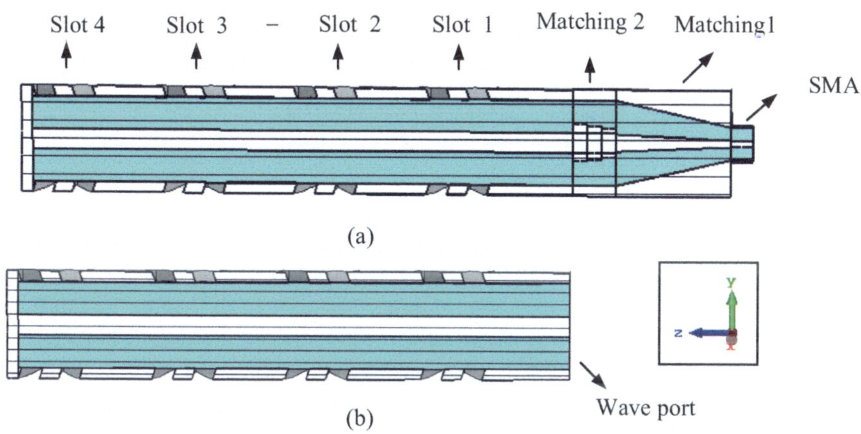

Fig. 8.12 The model of the reconfiguration pattern antenna. **a** With matching line. **b** Without matching line

second, transferring the standing wave to traveling wave to improve the performance of impedance matching.

Figure 8.13 shows the coaxial slot antenna fed by wave port. However, because the size of the coaxial line and the SMA connector do not match, and the impedance matching is poor. Therefore, a coaxial matching Sect. 8.2 with a gradient coaxial inner conductor radius and a coaxial matching Sect. 8.1 are added. The size of the coaxial matching Sect. 8.1 matches the SMA connector. The coaxial matching Sects. 8.1 and 8.2 have a certain impedance matching function. Figure 8.12 is a schematic diagram

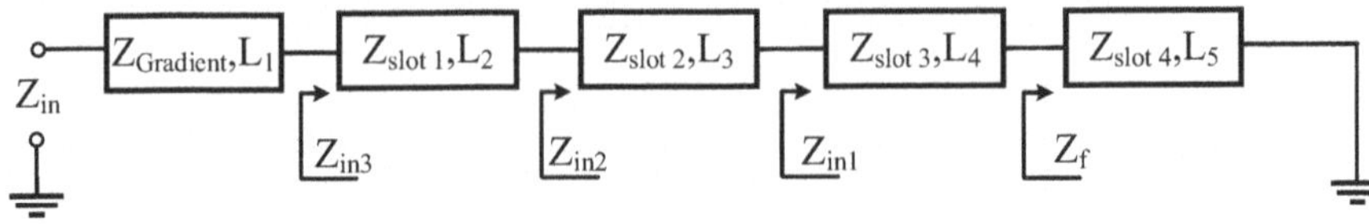

Fig. 8.13 The equivalent circuit of the antenna

of an equivalent circuit of the antenna loaded with a coaxial matching section. The slot at the extreme end of the coaxial line is equivalent to loading a certain complex impedance Z_f. In the edn, the impedance is change to Z_{in}, In general, the input impedance of the antenna is 50. Based on the basic theory of transmission lines, its input impedance is:

$$Z_{in1} = Z_{c1}\frac{Z_f + jZ_{c1}\tan\beta L_4}{Z_{c1} + jZ_f\tan\beta L_4} \tag{8.4}$$

$$Z_{in2} = Z_{c1}\frac{Z_{in1} + jZ_{c1}\tan\beta L_4}{Z_{c1} + jZ_{in1}\tan\beta L_4} \tag{8.5}$$

$$Z_{in3} = Z_{c1}\frac{Z_{in2} + jZ_{c1}\tan\beta L_4}{Z_{c1} + jZ_{in2}\tan\beta L_4} \tag{8.6}$$

$$Z_{in} = f_c\frac{Z_{in3} + jZ_{c1}\tan\beta L_4}{Z_{c1} + jZ_{in3}\tan\beta L_4} \tag{8.7}$$

where the characteristic impedance is $f_{c.}$

8.4 Summary

In this chapter, we have proposed compact reconfigurable coaxial antenna with novel slots containing switches. Switch and antenna is compact connection. The directional beams can be flexible controlled to cover entire horizonal plane. The realized gain in the operation bandwidth (11 GHz-11.5 GHz) is over 9 dBi, and, the maximum gain in the main lobe radiation direction is 10.2 dBi.

References

1. X.S. Yang, B.Z. Wang, W. Wu, S. Xiao, Yagi patch antenna with dual-band and pattern reconfigurable characteristics. IEEE Antennas Wireless Propag. Lett. **6**, 168–171 (2007)
2. M. Maqsood et al., Low-cost dual-band circularly polarized switched-beam array for global navigation satellite system. IEEE Trans. Antennas Propag. **62**(4), 1975–1982 (2014)

3. Chao Gu, S. Gao, Frequency-Agile beam-switchable antenna. IEEE Trans. Antennas Propga. **65**(8), 3819–3826 (2017)
4. A. Edalati, T.A. Denidni, High-gain reconfigurable sectoral antenna using an antive cylindrical FSS structure. IEEE Trans. Antennas Propga. **59**(7), 2464–2472 (2011)
5. B. Zhou, J.P. Geng, An omnidirectional circularly polarized slot antenna array with high gain in a wide bandwidth. IEEE Antennas and Wireless Propga **14**, 666–669 (2014)
6. Lei. Wang, et al., A Compact Reconfigurable coaxial slot antenna, 2018 International Applied Computational Electromagnetics Society Symposium-China (ACES), Beijing, China, 2018, pp. 1-2

Chapter 9
The Omnidirectional/Directional Switchable Antenna Based on the Curved Microstrip Antenna with Defected Ground Structure

9.1 Introduction

With the quickly developing of wireless transmission, the miniaturized terminal with multi-function working in multi-frequency bands has attracted more and more users. However, with the digital circuit being packaged into chip, the terminal miniaturization has gone to bottleneck period. The microwave transmission line and antenna is difficult to reduce size, but the RF module and antenna cover more than 80% surface of the terminal and more than 60% size of the terminal. Miniaturize the dimension of the microwave circuit and antenna has been critical problem. Usually, people hope to integrate the antenna inside the terminal, and it can work in several available frequency bands with being easy conformal, high gain and expected far field. Such as the external whip antenna, it is very convenient for WLAN terminal over 2.4 GHz, 5.25 GHz or 5.8 GHz, but it is not pretty and easily to be damaged. On the contrary, microstrip antenna, slot antenna and their array are easily manufactured with low cost and can be used in WLAN system [1–6] for good performances with low profile, lightweight. Especially, they are easy to integrated with devices.

Conformal antenna, especially, the curved microstrip antenna being conformal with the cylinder supporter has cause widespread concerns by many researchers [7–9], which can widen the beam width and decreasing the antenna size.

Electromagnetic bandgap (EBG) or photonic bandgap (PBG) structure has good band gap property, and is very useful to reduce the surface waves [10–16]. It has been inserted under the patch to suppress the surface wave along the ground and raise the gain of antenna [17–19]. But it is difficult to implement in large area.

Similar as EBG, the defected ground structure (DGS) is another 2D microwave structure, it can adjust the guided wave with bandpass and bandstop properties like filter. It is easy to realize by cutting special form in the ground, and change the current distribution on the ground so that change the equivalent impedance and the performance of the antenna. DGS is simple, and needn't periodically cut in larger area [20]. DGS has been widely used to improve the characters of the microstrip antenna [21–23] such as miniaturization, widening bandwidth and gain enhancement.

J. Geng, *Omnidirectional Slots Antenna*,
https://doi.org/10.1007/978-981-15-9089-4_9

In [24], the DGS is introduced to design the compact, wide band microstrip antenna, and the realized available bandwidth is about 4.3 times that of the traditional microstrip antenna. In [22], the DGS of several slots are cut in the ground of the microstrip antenna, it is true that achieved the compact dimension, wide band and high gain. In [25], the DGS in the slot-load to the microstrip slot antenna help to realize two wider bands and being lower band towards. In further, DGS can remove the higher order harmonics, improve return loss and reduce the size of antenna.

In this chapter, the DGS is introduced in the ground to realize the useful multiband, small size and high gain. The series of curved cylindrical antennas with DGS in the ground are proposed to work in multiband. The designed cylindrical curved conformal antennas are more smaller with wider pattern beam, they are suitable for WLAN system to be fit for different environment. At First, the original protype of the planar microstrip antennas with DGS is presented. The protype by optimizing shows good performance for the triple frequency 2.45, 5.25, and 5.8 GHz. Then a series of cylindrical conformal antennae are curved with different curving angle. The characters of these antennas are studied and the principle of the curving angle to the conformal antennas are analyzed exploring the simulations and measured results.

9.2 Geometry Structure and Analysis Method

9.2.1 Antenna Geometry with Defected Ground

The original protype is a rectangular patch antenna as illustrated in Fig. 9.1a. The rectangular patch is in size: *length1* × *length2*, and several rectangular holes are cut in the ground in Fig. 9.1b, and the parameters are marked in the model.

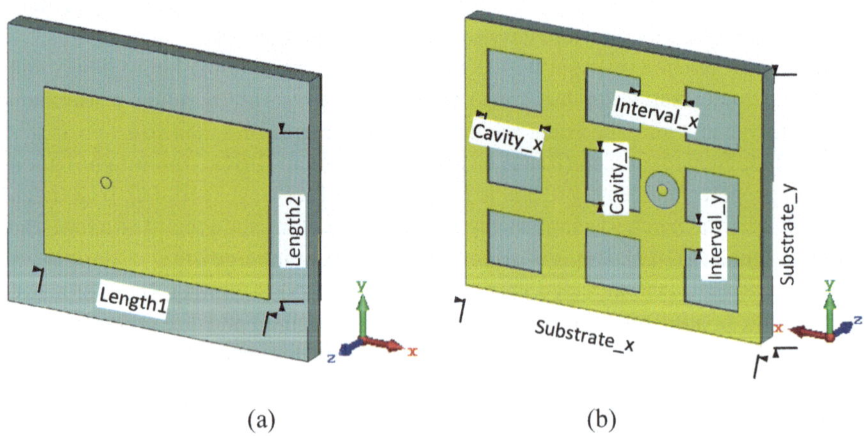

Fig. 9.1 The rectangular patch antenna with DGS: **a** Front side; **b** Back side

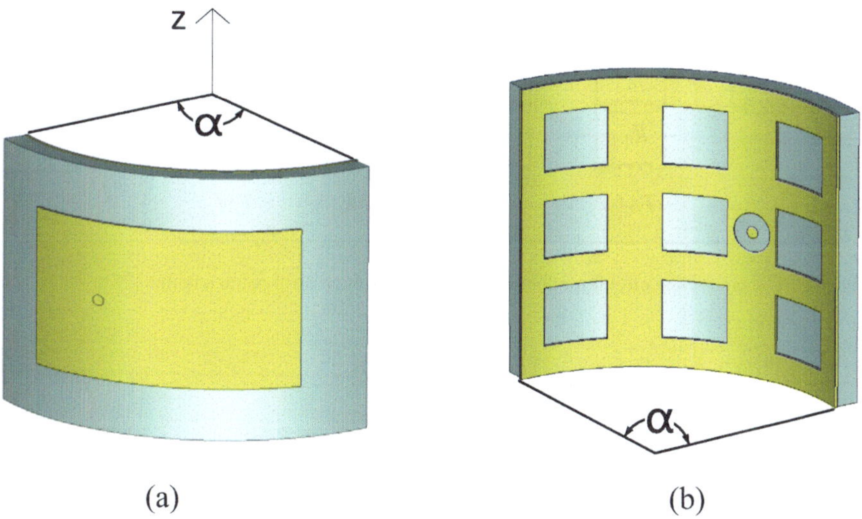

Fig. 9.2 The curved rectangular antenna, the curving angle is α. **a** Front side; **b** back side

Rectangular patch is dual-orthogonal-linear polarization antenna on the whole ground. But the antenna ground in Fig. 9.1b has been carved many holes as DGS, and the antenna patch is smaller than DGS ground, and it can work with two linear-orthogonal polarization modes too.

Similar as reported in [26], the DGS ground in Fig. 9.1b has frequency selective characters, and the size of DGS is related to the band stop and band pass frequency bands. And the work band of the patch antenna is filtered by the combination of the patch and the DGS ground.

Figure 9.2 shows the curved antenna with the curving angle α being conformal to cylinder with Z axis, in which the patch and DGS sizes are almost same as the antenna in Fig. 9.1.

9.2.2 *Equivalent Circuit*

Based on the cavity model theory, the rectangular microstrip antenna can be equivalent to a general lumped element circuit as shown in Fig. 9.3 by the mode expansion technique [27, 28].

In Fig. 9.3, the equivalent input impedance Zeq(f) of the patch antenna can be deduced as:

$$Z_{eq}(f) = j2\pi f L_{\infty} + \frac{1}{j2\pi f C_{0,0}} + \sum_{i=1}^{M} \frac{R_{0,i}}{1 + jQ_{0,i}(\frac{f}{f_{0,i}} - \frac{f_{0,i}}{f})} \tag{9.1}$$

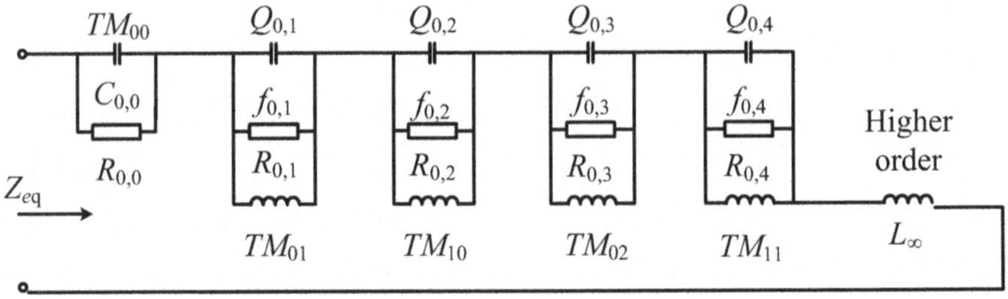

Fig. 9.3 The equivalent circuit model to the rectangular microstrip patch antenna [28]

If frequency $f \to 0$, Eq. (9.1) can be written as

$$\lim_{f\to 0} Z_{eq}(f) = \lim_{f\to 0} \frac{1}{j2\pi f C_{0,0}} + jf(2\pi L_\infty + \sum_{i=1}^{M} \frac{R_{0,i}}{Q_{0,i} f_{0,i}}) = \frac{1}{j2\pi f C_{0,0}} \tag{9.2}$$

Then, the capacitance $C_{0,0}$ of TM_{00} mode can be achieved by Eq. (9.3)

$$\begin{aligned} C_{00} &\approx \lim_{f\to 0} \frac{1}{2\pi f^2 \frac{\partial \mathrm{Im}[Zm(f)]}{\partial f}} \\ &\approx \frac{6\Delta f}{\pi f_3^2 \mathrm{Im}[Z_m(f_1) - 8Z_m(f_2) + 8Z_m(f_4) - Z_m(f_5)]} \end{aligned} \tag{9.3}$$

In another side, if frequency $f \to \infty$, the equivalent input impedance Zeq(f) of Eq. (9.1) can be writened as

$$\lim_{f\to\infty} Z_{eq}(f) = j2\pi L_\infty f + \frac{1}{jf}(\frac{1}{2\pi C_{0,0}} + \sum_{i=1}^{M} \frac{R_{0,i} f_{0,i}}{Q_{0,i}}) = j2\pi L_\infty f \tag{9.4}$$

So, the inductance L_∞ can be approximated as

$$\begin{aligned} L_\infty &\approx \lim_{f\to\infty} \frac{\partial \mathrm{Im}[Z_m(f)]}{2\pi \partial f} \\ &\approx \frac{\mathrm{Im}[Z_m(f_{N-4}) - 8Z_m(f_{N-3}) + 8Z_m(f_{N-1}) - Z_m(f_N)]}{24\pi \Delta f} \end{aligned} \tag{9.5}$$

$$R_{0,i} \approx \mathrm{Re}[Z_m(f)]|_{f=f_{0,i}} \tag{9.6}$$

$$Q_{0,i} \approx -\frac{f_{0,i}}{2R_{0,i}} \times \mathrm{Im}[\frac{\partial Z_m(f)}{\partial f}]|_{f=f_{0,i}} \tag{9.7}$$

There will be the introduced impedance around the feeding part when the microstrip antenna is fed by the probe, and it could be written as [29]

$$X_L = \frac{377h}{\lambda_0} \ln \frac{\lambda_0}{\pi d_0 \sqrt{\varepsilon_r}} \tag{9.8}$$

(a) Microstrip antenna with DGS

When the DGS is used in the planar microstrip antenna, there are advantages of size reduction, impedance band and gain enhancement [21, 24, 25]. So, we reviewed some merits for DGS embedded in the ground:

(i) The DGS is equivalent to a larger capacitance, so that decrease the antenna size in resonance.
(ii) DGS has the slow-wave character, and it can be equivalent to additional capacitance and inductance [30].
(iii) The rectangular holes in the ground means additional current length, and being equivalent to additional inductance.
(iv) Although being different the traditional microstrip antenna, the antenna with DGS can be regarded as a serial parallel resonant modes circuit from the cavity mode theory.
(v) For frequency move to high, the DGS work in high index mode, and is regarded as capacitance parallelly connected with inductance to be resonant in a new frequency band to extend the original band [31].
(vi) The DGS can suppress the surface wave to increase the antenna gain.

So, the planar microstrip antenna inserted the DGS in the ground can be equivalent to the microwave circuit with several resonant branches in Fig. 9.3.

(b) Curved microstrip antenna with DGS

When the curved radius become small, the equivalent circuit is still similar as the above.

When the curved radius become great, the equivalent capacitance and inductance may be affected by the current distribution, but the main principle is similar as the above.

9.2.3 Simulation Method and Experiment Condition

At beginning, the TRL model and cavity model are used to analyze microstrip antenna. Later, Green function, moment method and finite element method are used to EM computation to antenna. Finite difference in time domain (FDTD) method and Finite Integration in Time domain (FIT) are used to full wave analyze too. Here, the commercial EM software CST MICROWAVE STUDIO based on FIT and FDTD is used to simulate the conformal curved antenna [32, 33].

The Agilent vector network analyzer 8722ES is adopted to measure the impedance match bandwidth of the antenna. The far fields and gain of the designed antenna are measured on the top of our lab building, the a half open space The transmitter source is the signal generator Agilent PSG E8267D, and the spectrum analyzer Agilent PSA E4447A is the receiver connecting with the designed antenna, which rotates in azimuth plane on the rotator. The gain is achieved estimated by comparing the experiment maximum power level data from the designed antenna and the maximum value received from the standard horn antennas at the same frequencies [34].

9.3 Result Analysis for Planar Antenna with DGS

The original parameters value of the microstrip antenna protype are: permittivity of the substrate ε_r is 2.02, and the thickness of the substrate layer is 2 mm, copper layer thickness is 0.018 mm, interval_u = interval_v = 3 mm, cavity_u = 10.8 mm, cavity_v = 11.7 mm, substrate_u = 45 mm, substrate_v = 47 mm, length1 = 35.5 mm, length2 = 39 mm. The antenna is excited from the SMA adaptor with characteristic impedance 50 O.

9.3.1 Impedance

For the planar microstrip antennas without DGS and another one with DGS are constructed, their external parameter values are same as the above, and the calculated input impedance (real part and imaginary part) of these two antenna are shown in Fig. 9.4.

It can be seen that the impedance-frequency curves for the mircrostrip antenna with DGS are similar with those of the antenna without DGS, and the resonant frequency move to lower band clearly, which is in consistent of the calculation results from the equivalent circuit of the antenna with DGS.

Based on the cavity model, the parameters of the equivalent circuit on different resonant models are calculated and shown in the Table 9.1. And the working models of the microstrip antenna without DGS and another one with DGS are similar, which means the cavity model still useful to the microstrip antenna with DGS. It can be seen that the working models TM01, TM10, TM02 and TM11 of the microstrip antenna with DGS all move to the lower band, C_{00} decreases, L_∞ increases, and their Q_{0i} becomes smaller too.

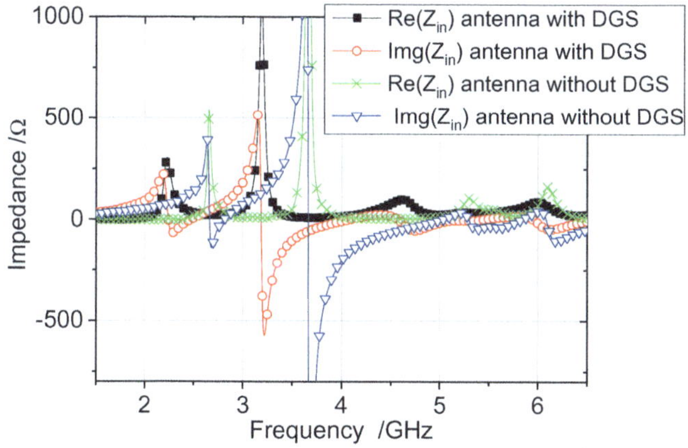

Fig. 9.4 Input Impedance: real part and Imaginary part for same external size microstrip antenna without DGS and another one with DGS (reproduced courtesy of The Electromagnetics Academy)

Table 9.1 The parameters of these working models for patch antenna without DGS and patch antenna with DGS

		Patch (no DGS)	Patch (with DGS)
	C_{00} (pF)	17.08 pF	12.07 pF
	L_{∞} (nH)	8.27 nH	13.45 nH
TM01	f_{01} (GHz)	2.664	2.254 GHz
	R_{01} (Ω)	459.64	236.06
	Q_{01}	43.38	12.26
TM10	f_{02} (GHz)	3.675 GHz	3.199 GHz
	R_{02} (Ω)	2609.0	1108.3
	Q_{02}	52.24	46.10
TM02	f_{03} (GHz)	5.306 GHz	4.578 GHz
	R_{03} (Ω)	108.72	98.28
	Q_{03}	24.85	9.21
TM11	f_{04} (GHz)	6.093 GHz	5.957
	R_{04} (Ω)	154.18	81.35
	Q_{04}	28.77	8.09

It is noted that *length1*, *length2* and *cavity_u* are key parameters, which affect the simulation results sensitively. And these parameters are analyzed in the follow in detail.

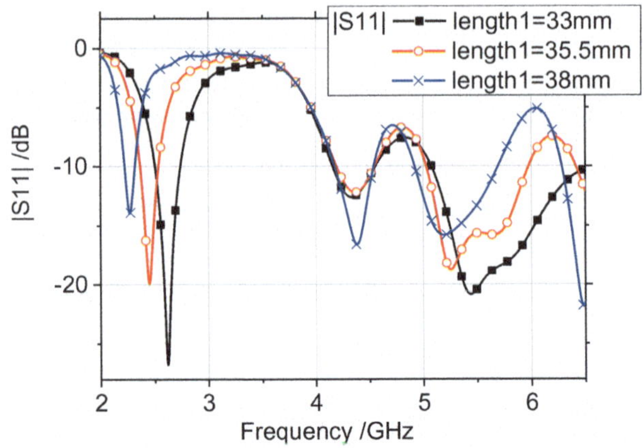

Fig. 9.5 The length1's impact on all the resonance frequencies (reproduced courtesy of The Electromagnetics Academy)

9.3.2 *Length1*

In Fig. 9.5, the return loss curves at different values of *length*1 in the planar microstrip antenna with DGS are illustrated. It can be seen that the pits in the reflection coefficient curves move to lower band around 2.45, 5.25, and 5.8 GHz with *length*1 increasing from 33 mm to 38 mm.

The details of the working modes changing by the variance of *length*1 are shown in Table 9.2. With *length*1 increasing, the inductance L_{∞} rises rapidly, and 4 working models TM01, TM10, TM02 and TM11 all are affected greatly, and move to lower bands.

9.3.3 *Length2*

In Fig. 9.6, the return loss curves at different values of *length*2 in the planar microstrip antenna with DGS are given. It can be seen that the troughs in the |S11| curves keep unmoved near 2.45 GHz, appear turning point near 5.25 GHz, but shallow near 5.8 GHz with *length*2 increasing from 37 mm to 41 mm.

The details of the working modes changing by the varying *length*2 are shown in Table 9.3. With *length*2 increasing, the capacitance C_{00} increase little, the inductance L_{∞} rises distinctly, and 4 working models TM01, TM10 and TM02 almost are unchanged, but TM11 varies greatly (Fig. 9.7).

Table 9.2 The parameters of these 4 models for patch antenna with DGS with length1 varying

Length1(mm)		33	35.5	38
C_{00}(pF)		10.69	11.03	12.13
L_{∞}(nH)		9.03	17.54	20.48
TM01	f_{01} (GHz)	2.338	2.24	2.121
	R_{01} (Ω)	270.45	276.00	306.4329
	Q_{01}	11.07	14.31	21.76
TM10	f_{02} (GHz)	3.283	3.178	3.069
	R_{02} (Ω)	562.15	1253.55	1960.62
	Q_{02}	23	48.82	72.55
TM02	f_{03} (GHz)	4.606	4.592	4.55
	R_{03} (Ω)	102.88	104.24	100.55
	Q_{03}	8.9	9.61	11.79
TM11	f_{04} (GHz)	6.0836	5.943	5.754
	R_{04} (Ω)	74.29	91.65	107.63
	Q_{04}	6.82	10.24	11.28

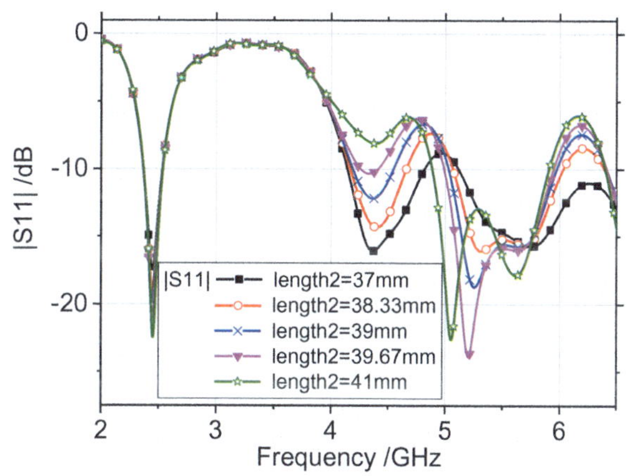

Fig. 9.6 The length2's impact on all the resonance frequencies (reproduced courtesy of The Electromagnetics Academy)

9.3.4 *Cavity_u*

In Fig. 9.6, the return loss curves at different values of *cavity_u* in the planar microstrip antenna with DGS are illustrated. It can be seen that the troughs in the |S11| curves become deeper around 2.45 and 5.8 GHz, but shallow around 5.25 GHz with *cavity_u* increasing from 9.8 mm to 11.8 mm.

The details of the working modes changing by the varying *cavity_u* are shown in Table 9.4. With *cavity_u* increasing from 9.8 mm to 11.8 mm, the capacitance C_{00}

Table 9.3 The parameters of these 4 models for patch antenna with DGS with length2 varying

Length2 (mm)		37 mm	38.33	39	39.67	41
C_{00} (pF)		10.64	10.89	11.03	11.17	11.67
L_{∞} (nH)		13.32	16.22	17.54	18.67	20.21
TM01	f_{01} (GHz)	2.24	2.24	2.24	2.24	2.247
	R_{01} (Ω)	276.83	273.43	276.00	277.72	272.03
	Q_{01}	14.86	14.24	14.31	14.46	13.68
TM10	f_{02} (GHz)	3.22	3.195	3.178	3.164	3.129
	R_{02} (Ω)	1268.26	1261.2	1253.55	1227.74	1196.35
	Q_{02}	49.11	52.74	48.82	47.83	49.48
TM02	f_{03} (GHz)	4.494	4.592	4.592	4.575	4.49
	R_{03} (Ω)	71.65	87.3	104.24	117.94	121.56
	Q_{03}	4.39	5.92	9.61	11.91	10.23
TM11	f_{04} (GHz)	6.027	5.971	5.943	5.926	5.894
	R_{04} (Ω)	79.45	89.27	91.65	93.2	94.34
	Q_{04}	8.05	10.18	10.24	10.5	11.21

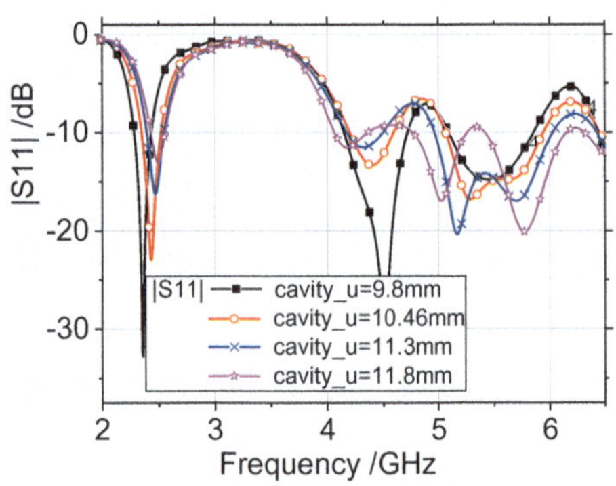

Fig. 9.7 The cavity_u's impact on all the resonance frequencies (reproduced courtesy of The Electromagnetics Academy)

almost keep unmoved, the inductance L_{∞} decreases, and 4 working models TM01, TM10, TM02 and TM11 all varied.

9.3.5 *The Optimized Parameter Values for the Plane Microstrip Antenna with DGS*

With the above analysis, the final parameters of the antenna are optimized and given as following, permittivity of the substrate ε_r is 2.02, and the thickness of the substrate layer is 2 mm, copper layer thickness is 0.018 mm, *interval_u* = *interval_v* = 3 mm, cavity_u = 10.8 mm, *cavity_v* = 11.8 mm, *substrate_u* = 45 mm, *substrate_v* =

Table 9.4 The parameters of these 4 models for patch antenna with DGS with *cavity_u* varying

cavity_u(mm)		9.8	10.46	11.13	11.8
C_{00}(pF)		11.99	10.96	11.23	10.83
L_{∞} (nH)		19.28	17.08	16.24	15.22
TM01	f_{01} (GHz)	2.219	2.236	2.254	2.252
	R_{01} (Ω)	175.36	263.99	313.11	419.03
	Q_{01}	8.17	13.72	16.54	23.46
TM10	f_{02} (GHz)	3.115	3.157	3.199	3.227
	R_{02} (Ω)	1329.46	1340.27	1138.45	1240.18
	Q_{02}	42.75	51.28	45.68	51.46
TM02	f_{03} (GHz)	4.697	4.62	4.55	4.403
	R_{03} (Ω)	89.94	100.83	102.96	99.01
	Q_{03}	10.50	9.54	8.4	8.09
TM11	f_{04} (GHz)	5.936	5.964	5.971	5.999
	R_{04} (Ω)	112.19	100.05	87.53	76.29
	Q_{04}	12.82	12.39	10.58	10.16

47 mm, *length*1 = 35.5 mm, *length*2 = 39 mm.The antenna is fed by probe with 50 Ohm. The finally fabricated planar microstrip antenna with DGS is shown in Fig. 9.8 ($\alpha = 0°$). The measured reflection coefficients |S11| of the manufactured antenna compared with the simulation results are given in Fig. 9.10 (center angle $\alpha = 0°$).

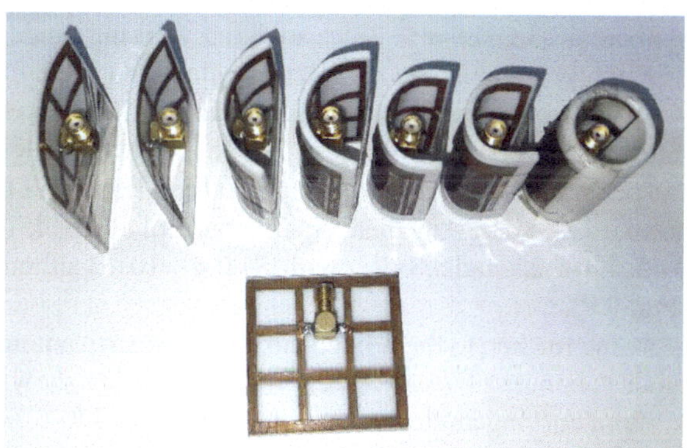

Fig. 9.8 Curved antennas with different central angle α (0°, 45°, 90°, 135°, 180°, 225° and 270°) for experiment

9.4 Curved Antennae

Based on the planar microstrip antenna with DGS in Fig. 9.8, the antenna is curved with various central angle α, which can be conformal to cylinder surface with compacted size of the antenna. Based on the curved model in Fig. 9.2, the microstrip antenna is curved with several different curving angle α around the Z axis, and these curving angle α is 0°, 45°, 90°, 135°, 180°, 225° and 270° in turn. The manufactured series of the curved microstrip antennas are presented in Fig. 9.8. And these antennas are studied by comparing the simulation and experiments results, and analyzed by approximated cavity model.

9.4.1 The Varying Principle of the Curved Antenna with Curving Angle α Changing, but Other Parameters Being Unmoved

Keeping the size parameters' value as the above, these curved microstrip antennas with different central angle α are simulated, and the return loss |S11| curves are shown in Fig. 9.10a. It can be seen that the reflection coefficient |S11| curves keep unmoved at 2.45 GHz, but raise in 5.25–5.8 GHz with the curving angle α increasing from 0° to 270°.

The details of the working modes changing by varying curving angle α are shown in Table 9.5. With curving angle α increasing from 0° to 270°, the capacitance C_{00} decrease a little, the inductance L_{∞} increases. With the curving angle α raising, the resonant frequency f_{03} of TM_{02} moves toward lower band, and a new mode appears at about 4.33 GHz, $\alpha = 225°$, so that 5 working modes coexist during 2–6 GHz.

Slightly refine some parameters value, and these curved antennas are fabricated according to the finally optimized parameter's values in Table 9.2. And the measured return loss |S11| curves of these curved microstrip antenna are shown in Fig. 9.10b. It can be seen that the measured reflection coefficient |S11| curves almost keep unmoved around 2.45 GHz. And there are four parts of available impedance bandwidths around 2.45, 4.4, 5.25, and 5.8 GHz with $|S11| < -10$ dB although these curves differs clearly (Fig. 9.9).

In the further, to the refined curved microstrip antenna, The details of the working modes changing by varying curving angle α are shown in Table 9.5.

With curving angle α increasing from 0° to 270°, 5 working models are excited in these antennas in 2–6 GHz, and the capacitance C_{00} decrease a little, the inductance L_{∞} increases, the TM01, TM10, TM02, TM11 and TM20 all are changed.

In fact, the parameters values of these curved microstrip antennas changed a little when the original planar microstrip antennas are curved in varied central angle α. Suppose the arc length of the curved antenna is *l*, the thickness of the planar microstrip antenna is *hd*. Keeping *l* and α (degree), then the inside length become smaller with the inside length error $\Delta l \leq 0$,

Table 9.5 The parameters of the working models for the curved patch antenna with DGS with the curving angle α varying

	α (°)	Curved antenna						Refined curved antenna					
		45°	90°	135°	180°	225°	270°	45°	90°	135°	180°	225°	270°
	C_{00} (pF)	11.68	11.63	11.45	11.3	11.25	11.16	11.81	11.76	11.73	11.76	11.19	11.36
	L_{∞} (nH)	17.21	15.67	16.67	15.88	15.98	8.313	23.14	20.07	24.78	35.98	42.8	23.40
TM01	f_{01} (GHz)	2.24	2.247	2.254	2.282	2.205	2.205	2.24	2.219	2.198	2.107	2.58	2.114
	R_{01} (Ω)	200.13	170.98	142.61	116.33	151.85	130.45	198.38	170.53	135.55	64.62	151.37	162.78
	Q_{01}	10.32	6.32	5.16	2.73	9.3	7.33	9.11	6.76	4.4	23.24	10.32	9.71
TM10	f_{02} (GHz)	3.22	3.234	3.248	3.255	3.262	3.283	3.199	3.15	3.132	3.115	3.101	3.115
	R_{02} (Ω)	1437.29	1197.64	1030.7	916.21	769.42	911.69	1389.46	1163.83	820.51	856.33	728.01	712.22
	Q_{02}	48.2	39.4	32.58	26.72	22.24	29.3	50.39	38.93	25.14	26.63	23.05	22.37
TM02	f_{03} (GHz)	4.564	4.585	4.704	4.837	4.333	4.375	4.55	4.515	4.557	4.641	4.032	4.543
	R_{03} (Ω)	109.37	113.28	148.16	307.55	30.83	35.25	147.18	252.73	354.94	253.99	29.99	205.15
	Q_{03}	7.61	7.96	14.04	44.69	8e−6	2.25	13.89	30.28	49.86	38.57	0.0005	33.80
TM11	f_{04} (GHz)	–	–	–	–	4.949	5.033	5.173	5.082	5.005	4.963	4.963	5.082
	R_{04} (Ω)	–	–	–	–	490.96	379.70	42.94	54.86	68.08	133.63	495.38	1118.20
	Q_{04}	–	–	–	–	72.04	52.15	2.813	6.12	6.37	19.06	62.98	136.7
TM20	f_{05} (GHz)	5.87	5.894	5.999	6.111	6.265	6.398	5.803	5.831	5.789	5.775	5.838	5.873
	R_{05} (Ω)	71.86	56.54	49.89	44.49	37.34	28.26	705	78.15	68.43	51.96	37.82	45.34
	Q_{05}	5.84	3.47	5.45	9.183	9.55	4.34	3.75	13.31	11.27	12.15	9.06	14.93

Fig. 9.9 |S11| with different curving angle α: **a** Simulation for curved antenna with original size; **b** measured |S11| for the curved antenna with refined parameter value (reproduced courtesy of The Electromagnetics Academy)

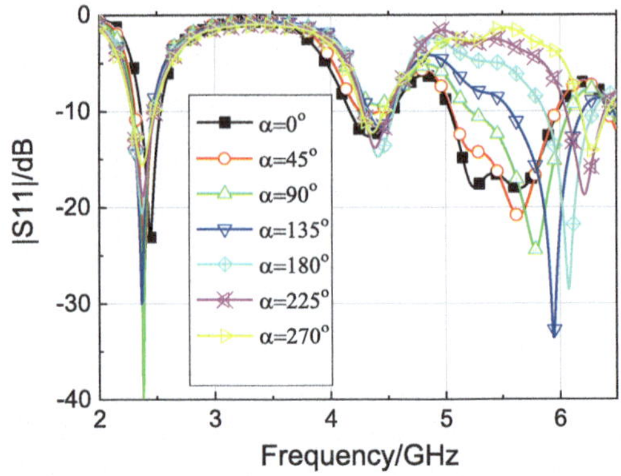

(a)

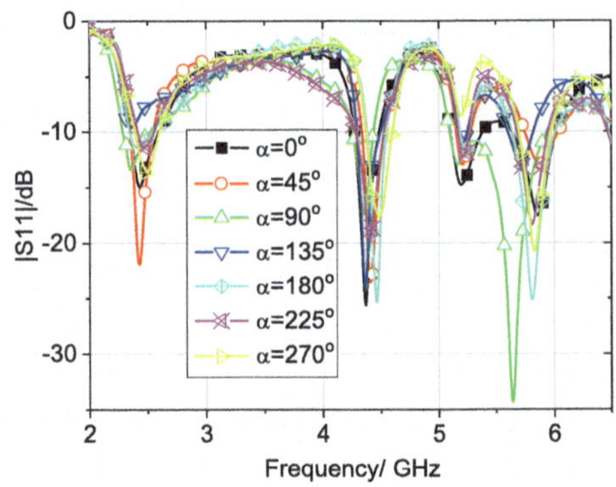

(b)

$$\Delta l = -hd \cdot \alpha \cdot \pi/180 \leq 0 \tag{9.9}$$

This's just the reason that the available frequency band of 5.2–5.8 GHz moves to higher frequency with α increasing in Fig. 9.10a.

Refer to results of length2 effect to impedance bandwidth in Fig. 9.6 in Sect. 9.3, the available frequency band of 5.2–5.8 GHz will move to lower frequency band with lenth2 increasing. So, length2 become larger in Table 9.4 with central curved angle α increasing to keep the available impedance frequency band into 5.2–5.8 GHz (in Fig. 9.10b).

Certainly, other parameters are adjusted slightly in Table 9.6 to better the impedance match too.

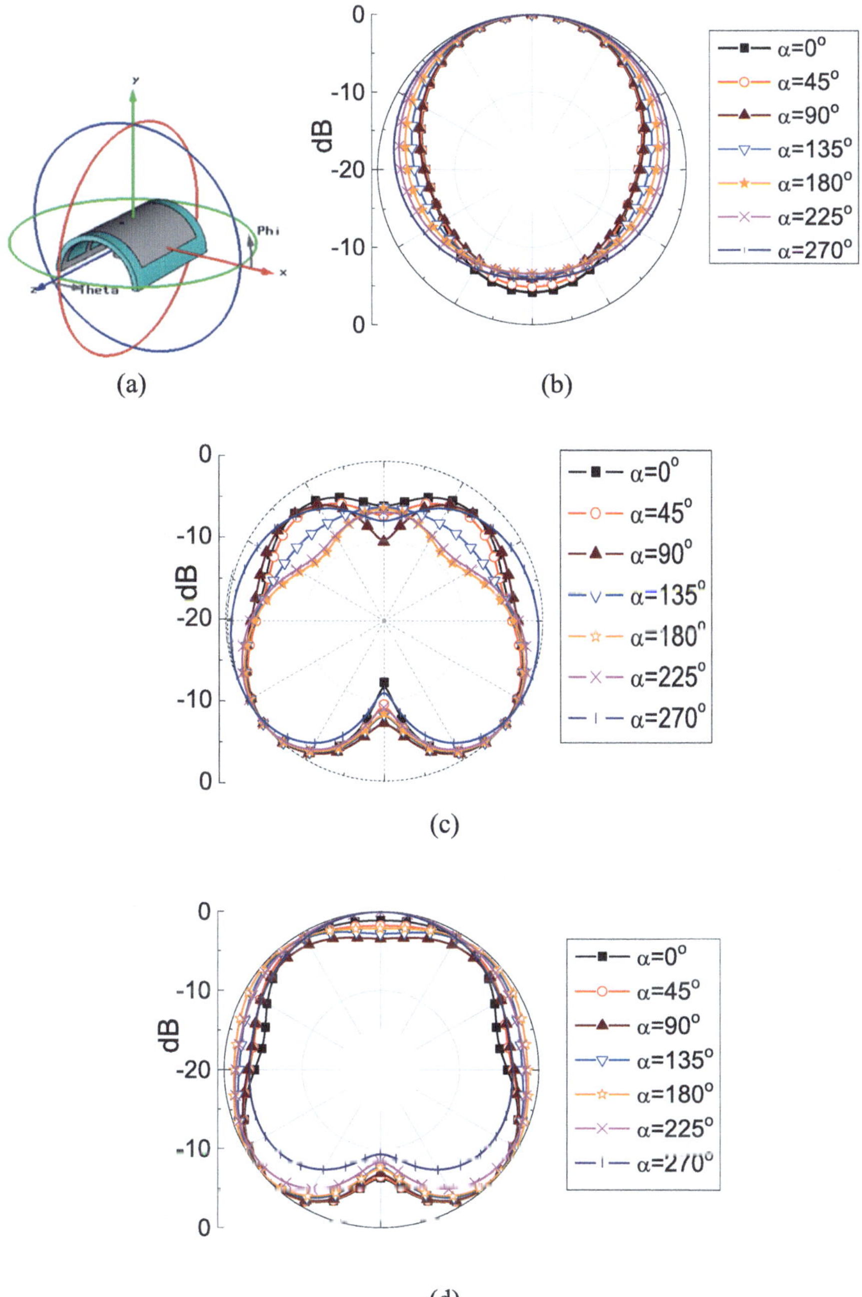

Fig. 9.10 Curved conformal antennas' simulation pattern in XY plane for the different curving angle α: **a** Coordinate system; **b** f = 2.45 GHz; **c** f = 5.25 GHz; **d** f = 5.8 GHz (reproduced courtesy of The Electromagnetics Academy)

Table 9.6 Modified parameters of the curved antennas with different central angles

Angle α (°)	0	45	90	135	180	225	270
cavity_u (mm)	10.8	10.8	10.8	10.8	10.8	10.8	10.8
cavity_v (mm)	11.8	11.8	12	12.3	12.6	12.9	12
interval_u (mm)	3	3	3	3	3	3	3
interval_v (mm)	3	3	2.7	2.3	2.2	1.8	1.6
length1 (mm)	35.5	35.5	35.5	35.5	35.5	35.5	35.5
length2 (mm)	39	40	42	42.8	43	43.4	44
substrate_u (mm)	45	45	45	45	45	45	45
substrate_v (mm)	47	47	47	47	47	47	47

9.4.2 Patten Varying with the Curving Angle α

The coordinate of these series curved antenna is fixed in Fig. 9.10a.

According to the parameter's value in Table 9.3, the simulated far fields patterns at 2.45 GHz, 5.25 GHz and 5.8 GHz with different curving angle α are given in Fig. 9.10.

At 2.45 GHz, the far field patterns of these curved antenna are compared in Fig. 9.10b, the radiation beam in xoy plane become wider with α increasing, and half power beam width (HPBW) at α = 270° is more than 180°.

At 5.25 GHz, the far field patterns of these curved antennas are compared in Fig. 9.10c, the radiation beam in xoy plane become wider with α increasing, but there exist two null points in y direction and −y direction.

At 5.8 GHz, the far field patterns of these curved antennas are compared in Fig. 9.10d, the radiation beam in xoy plane become wider with α increasing, but there appear one null point in −y direction. And the HPBW is more than 180° at α = 270°.

All these manufactured curved microstrip antenna are measured in half open space on the top of our lab building, and the measured far fields patterns are given in Fig. 9.11. And the measured far fields of these series of antennas are consistent with the simulation results, besides some small differences. And the measured results again certificate that the radiation beams in xoy plane become wider with α increasing.

9.4.3 Gain

The measured gains of these curved antennae with different curving angle α are compared with the simulation results at 2.45, 5.25, and 5.8 GHz in Table 9.7 and the measured gains are during 2–6.3 dBi, and the changing trend is similar with the simulated results.

Fig. 9.11 Curved conformal antennas' measured pattern in XY plane for the different curving angle α: (a) f = 2.45 GHz; (b) f = 5.25 GHz; (c) f = 5.8 GHz

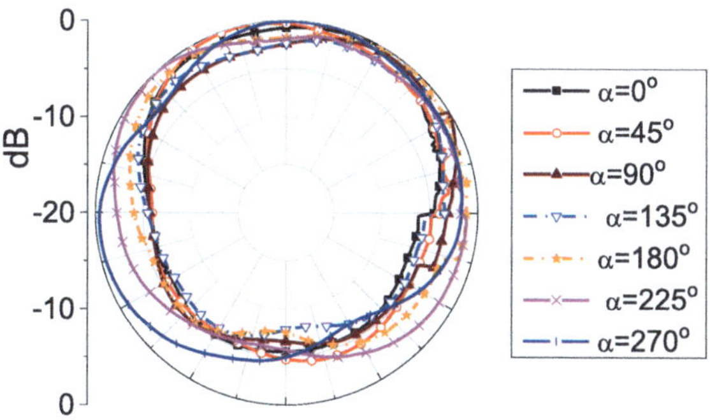

(a)

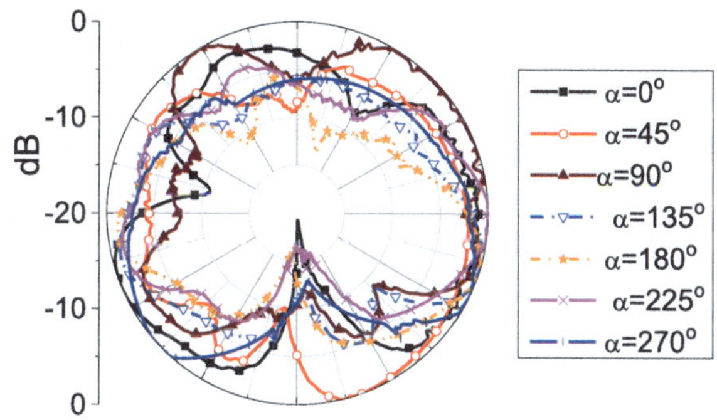

(b)

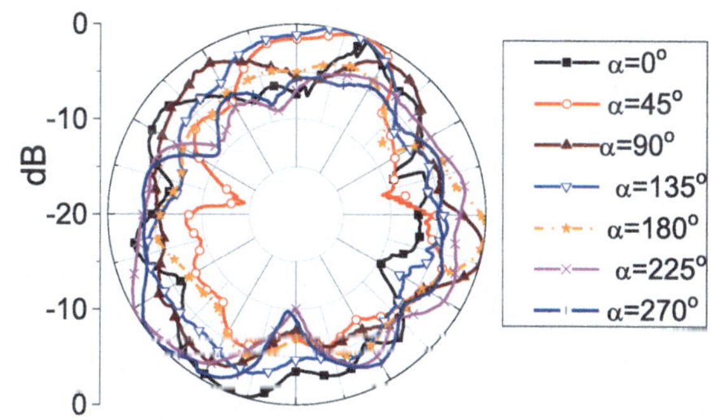

(c)

Table 9.7 Antenna gain

α (°)	f(GHz)					
	Simulated (dBi)			Measured (dBi)		
	2.45	5.25	5.8	2.45	5.25	5.8
0	5.224	5.129	2.84	5.815	4.485	3.583
45	5.501	4.106	3.297	4.905	4.568	2.577
90	5.421	5.297	3.934	6.305	3.843	3.826
135	5.306	3.509	4.423	5.075	4.376	4.027
180	4.993	3.311	3.744	5.375	3.038	2.206
225	4.631	3.314	3.666	4.815	3.242	3.718
270	4.254	3.467	3.852	3.805	4.342	4.804

9.5 Omnidirectional/Directional Switchable Antenna

9.5.1 Switchable Principle

In [35], as shown in Fig. 9.12, a balance switch diagram for omnidirectional or directional switchable antenna is illustrated. When the switch combines with the WLAN device, it can work in directional pattern mode or omnidirectional pattern mode by turning switches. It includes switch 1, switch 2, switch 3 and splitter/combiner, one RF input port 1 from the WLA device, another two ports (2 and 3) connecting to two direction antennas (Ant1 and Ant2) respectively. This pattern reconfigure system can work in several cases.

Case 1, only Ant1 works by switch 1 turning on AB, switch 2 on HK and switch 3 on MN. The radiation pattern is just like the pattern in Fig. 9.12b.

Case 2, only Ant2 works by switch 1 turning on AC, switch 2 on GK and switch 3 on MN. The radiation pattern is just like the pattern in Fig. 9.12c.

Case 3, Both Ant1 and Ant2 work together by switch 1 turning on AB, switch 2 on HK and switch 3 on LN. The radiation pattern is just like the pattern in Fig. 9.12d.

Case 4, no antenna works by switch 1 turning on AB, switch 2 on GK and switch 3 on MN, or by switch 1 turning on AC, switch 2 on GK or HK and switch 3 on LN.

In fact, switch 2 and switch 3 are just the dummy switches, which are introduced to keep the impedance match between case 1, case 2 and case 3.

Here, the selected RF switch chip is M/A-COM's MASWSS0070 as shown in Fig. 9.13. This chip is a broadband GaAs PHEMT MMIC SPDT switch available in a low cost 3 mm 12-lead PQFN package. The MASWSS0070 is ideally suited for application where very small size and low cost are required. Typical application is for WLAN 802.11b/g AP.

Then the RF switch system in Fig. 9.12 is simulated for different work mode. And the simulated performances of directional work mode are shown in Fig. 9.14. The work band of these ports cover 2–7 GHz, the insertion loss between port 1 and port

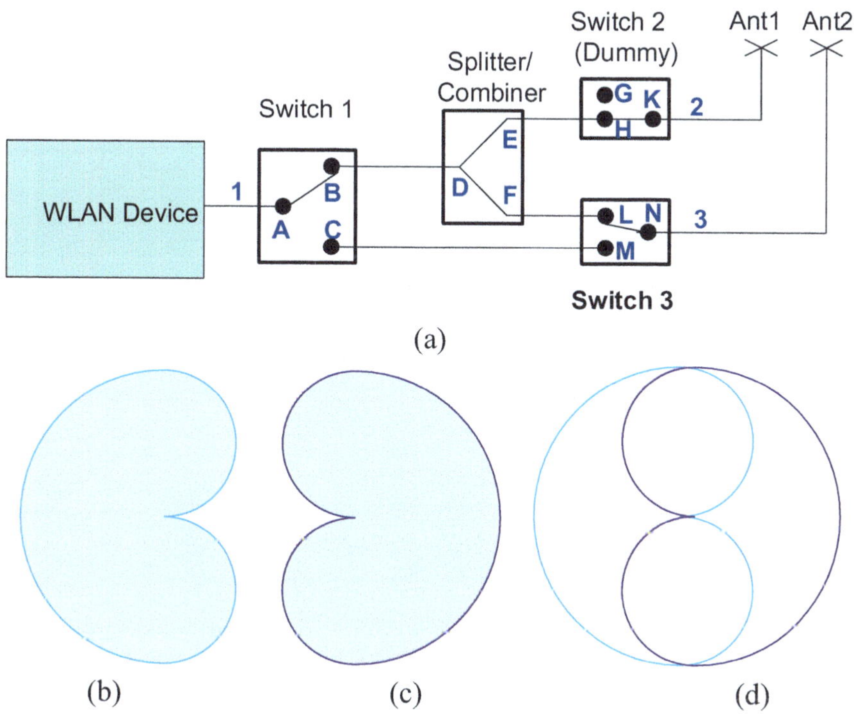

Fig. 9.12 Balance switch diagram with impedance match in three cases. **a** diagram, **b** case 1: Ant1 work, (2) case 2: Ant2 work, and **c** case 3: both Ant1 and Ant2 work together

Fig. 9.13 RF switch chip M/A-COM's MASWSS0070

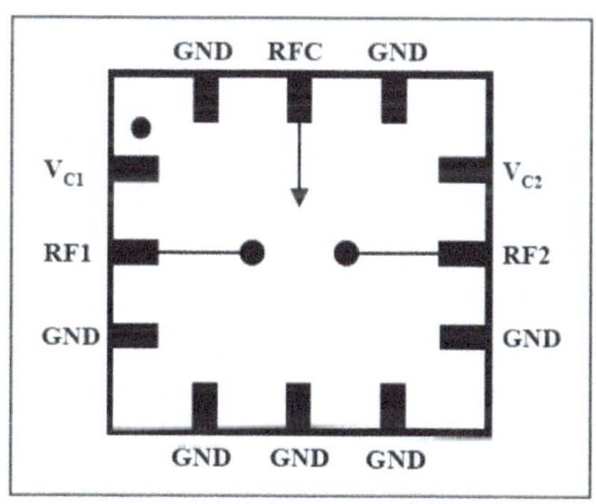

2 is very small only 0.12 dB, and the isolation is high enough between port 1 and port 3.

As discussed above, for case 3, the omnidirectional work mode, the RF switch system is simulated and the performance are given in Fig. 9.14. The available work band of |S11| < −10 dB is 2.25–7 GHz. And the transmission loss between port

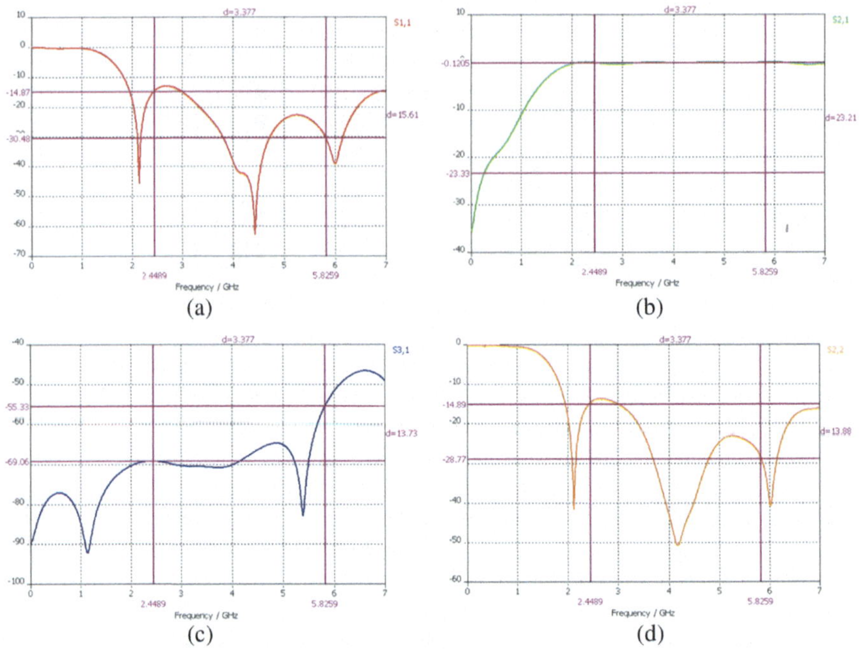

Fig. 9.14 Directional work mode, **a** S11, **b** S21, **c** S31 and **d** S22

1 and port 2 is about 0.13 dB, and almost same as that between port 1 and port 3 during 2.4–7 GHz. The available work band of |S22| < −10 dB is 2.15–7 GHz, but the available work band of |S33| < −10 dB is 1.3–7 GHz. So, the effective band of the system for these reconfigurable states is 2.4–7 GHz.

9.5.2 Omnidirectional/Directional Antenna

Based on the Curved antenna principle in 9.4 section, the curved patch with DGS is designed to directional pattern with wide HPBW [36]. And two elements are combined to realize space pattern diversities with the reconfigurabl52e circuit in Fig. 9.12. The antenna structure is shown in Fig. 9.15a, two curved antenna elements are piled one above another along same cylinder axis. The cut plane of the combined antenna is given in Fig. 9.15b. The curved patch is copper, the substrate is Teflon with permittivity 2.2, and the curved DGSs are on the inner wall of the Teflon cylinder. The whole radius of antenna is 8.24 mm, length 153 mm and the thickness of the Teflon substrate is 1 mm. Two feed points are in the DGS with SMA connector. The combined antenna is manufactured as shown in Fig. 9.1c.

The antenna is simulated, and the return loss curve is shown in Fig. 9.16. It is clear, the antenna can work at 2.45 GHz and 5.85 GHz.

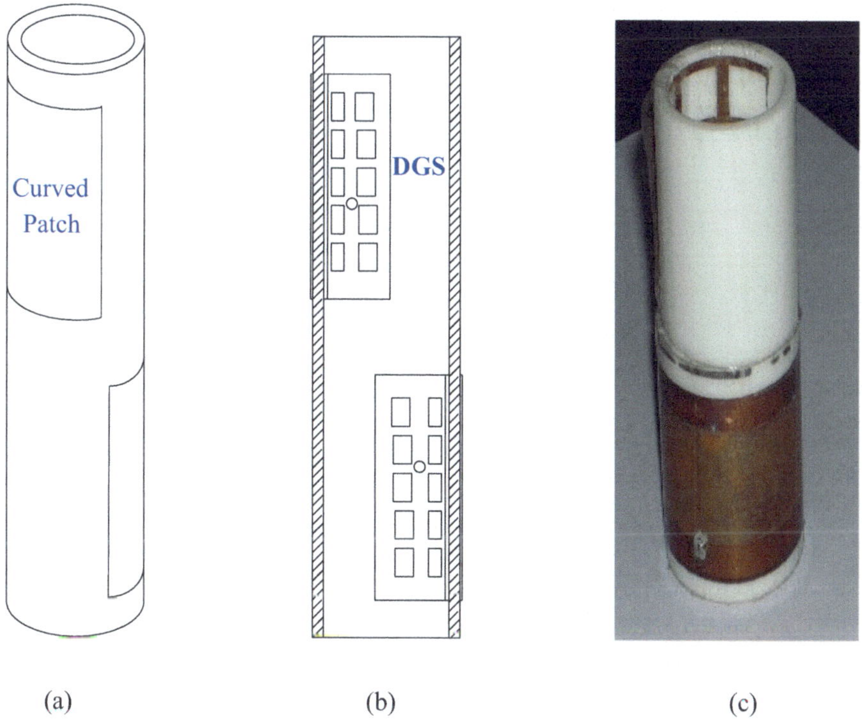

Fig. 9.15 Omnidirectional/directional antenna. **a** 3D structure of the antenna, **b** cut plane of the antenna, and **c** the manufactured antenna

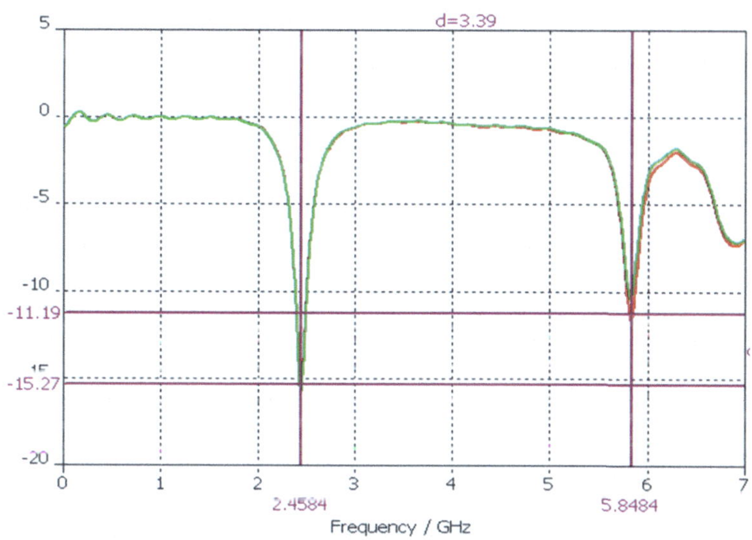

Fig. 9.16 Simulated return loss curve of the antenna

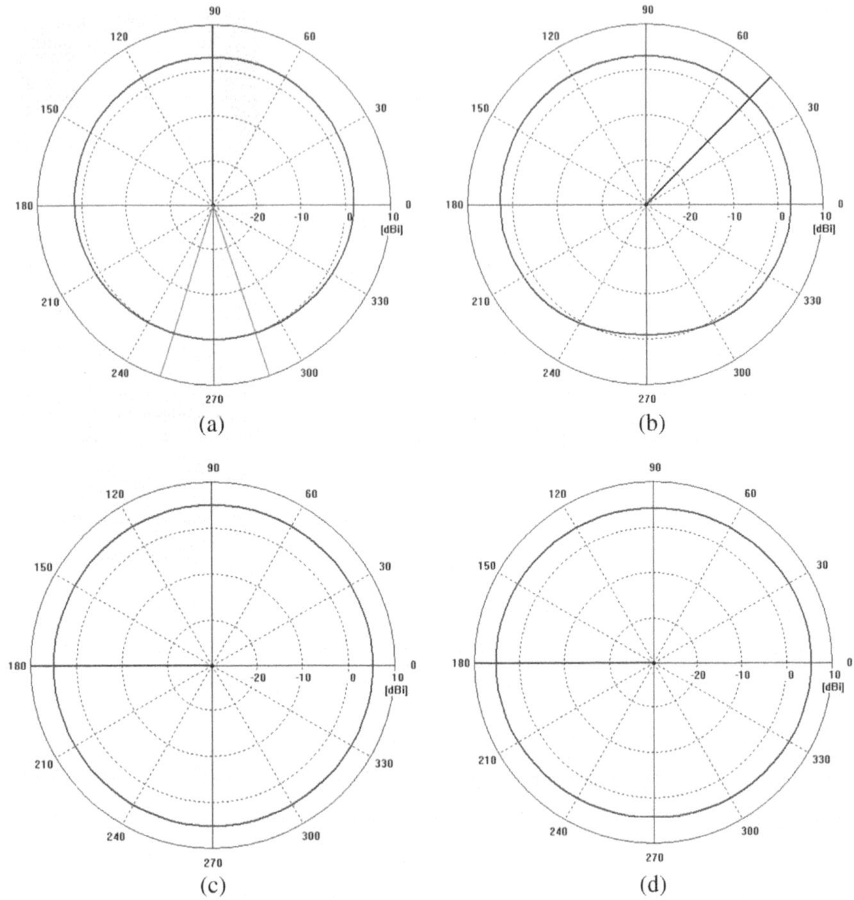

Fig. 9.17 The radiation patterns of the combination antenna. **a** Directional pattern at 2.4 GHz and **b** 5.8 GHz. **c** Omnidirectional pattern at 2.4 GHz and **d** 5.8 GHz

The radiation patterns of the combination antenna are shown in Fig. 9.17. The directional mode at f = 2.4 GHz is shown in Fig. 9.17a, and directional pattern at f = 5.8 GHz is in Fig. 9.17b. In Fig. 9.17c, d, the combined omnidirectional patterns are given.

9.6 Summary

The DGS is adopted in the ground design of the planar microstrip antenna to achieve the useful multiband, small size and gain enhancement, which is the basic of the curved conformal antenna, the relationship between the structure parameters and the antennas' characters are studied, which is the reference for the curved antenna.

With the curving angle α increasing, the return loss at f = 2.45 GHz almost keep good, but becomes poor at f = 5.25 GHz and 5.8 GHz. After slightly tuning the key parameters, these curved conformal antennas all can work well at f = 2.45 GHz, 5.25 GHz and 5.8 GHz, and their patterns become more omni-directional with the curving angle α increasing, which are verified by experiment too. The measured gain of these antennas are 2–6.3 dBi.

In further, the omnidirectional/directional switchable antenna system is introduced. The switch function is realized by a balance switch circuit with dummy switch to keep impedance match in different switch states. The antenna includes two curved patch with DGS, and piled one above another. When any one curved patch is excited, it radiates directional pattern. When both of the two curved patches are excited, they combine to the omnidirectional pattern.

References

1. Wei He, Ronghong Jin, Junping Geng, Mattias Lampe, Multiband dual patch antennas with polarization compensation for WLAN applications. Microwave Opt. Technol. Lett. **49**(8), 1907–1911 (2007)
2. He Wei, Jin Ronghong, Geng Junping, Gao Bo, Multiband antenna system with polarization conversion for wlan applications. Microwave Opt. Technol. Lett. **49**(7), 1772–1777 (2007)
3. R.K. Raj, M. Joseph, B. Paul, P. Mohanan, Compact planar multiband antenna for GPS, DCS, 2.4 = 5.8 GHz WLAN applications. Electr. Lett. **41**(6), 290–291 (2005)
4. Q. Zhong, Y. Li, H. Jiang, Y. Long, Design of a novel dual-frequency microstrip patch antenna for WLAN application, in *Antennas and Propagation Society International Symposium*, vol. 1 (IEEEJune 2004), pp. 277–280
5. Tuanjai Archevapanich'and Noppin Anantrasiric har', Inversed e-Shape Slot Antenna for WLAN Applica tions, in *International Conference on Control, Automation and Systems 2007*, Oct. 17–20, 2007 in COEX, Seoul, Korea, pp. 2854–2857
6. J. Janapsatya, K.P. Esselle, Multi-band WLAN antennas based on the principle of duality, in *Antennas and Propagation Society International Symposium 2006* (IEEE, 2006), pp. 2679–2682
7. K.-L. Wong, J.-S. Chen, Cavity-model analysis of a slot-coupled cylindrical-rectangular microstrip antenna. Microwaue Opt. Technol. Lett. **9**(20), 124–127 (1995)
8. Jin-Sen Chen, Kin-Lu Wong, Input impedance of a slot-coupled cylindrical-circular microstrip patch antenna. Microwaue Opt. Technol. Lett. **11**(1), 21–24 (1996)
9. J.S. Dahele, R.J. Mitchell, K.M. Luk, K.F. Lee, Effect of curvature on characteristics of rectangular patch antenna. Electron. Lett. **23**, 748–749 (1987)
10. E. Yablonovitch, Inhibited spontaneous emission in solid-state physics and electronics. Phys. Rev. Lett. **58**, 2059–2062 (1987)
11. D. Qu, L. Shafai, The performance of microstrip patch antennas over high impedance EBG substrates withinand outside its bandgap, in *2005 IEEE International Symposium on Microwave, Antenna, Propagation and EMC Technologies for Wireless Communications Proceedings* (2005), pp. 423–426
12. N. Jin, A. Yu, X. Zhang, An enhanced 2_2 antenna array based on a dumbbell EBG structure. Microwave Opt. Technol. Lett. **39**, 395–399 (2003)
13. W. He, R. Jin, J. Geng, Low RCS and high performances of microstrip antenna using fractal UC-EBG ground. IET Microwaves, Antennas Propag. **1**(5), 986–991 (2007)
14. Wei He, Ronghong Jin, Junping Geng, Guoming Yang, 2 × 2 array with UC-EBG ground for low RCS and high gain. Microwave Opt. Technol. Lett. **49**(6), 1418–1422 (2007)

15. E.B. Brown, C.D. Parker, E. Yablonovitch, Radiation properties of a planar antenna on a photonic-crystal substrate. J. Opt. Soc. Am. B **10**(2), 404 (1993)
16. V. Radistic, Y. Qian, R. Cocciloli et al., Novel 2-D photonic bandgap structure for microstrip lines. IEEE Microwave Guided Wave Lett. **8**(2), 69 (1998)
17. M.B. Matttew, B.B. John, O.E. Henry et al., Two dimentional photonic crystals fabry-perror resonators with lossy dielectrics. IEEE Trans. Microwave Theory Tech. **49**(11), 2085 (1999)
18. Y. Qian, Dan Sievenpiper, V. Raisic, et al., A novel approach for gain and bandwidth enhancement of patch antennas, in *RAWON'98 Proceedings* (1998), p. 221
19. X. Wang, Y. Hao, .S. Hall, Dual-band resonances of a patch antenna on UC-EBG substrate, in *Asia-Pacific Microwave Conference Proceedings*, vol. 1(2005), pp. 4–8
20. F.Y. Zulkifli, E.T. Rahardjo, D. Hartanto, Radiation properties enhancement of triangular patch microstrip antenna array using hexagonal defected ground structure. Prog. Electromagn. Res M **5**, 101–109 (2008)
21. X.-C. Lin, L.-T. Wang, A wideband CPW-fed patch antenna with defective ground plane, in *IEEE Antennas and Propagation Society International Symposium*, vol. 4 (2004), pp. 3717–3720
22. K.L. Wong, J.S. Kuo, T.W. Chiou, Compact microstrip antennas with slots loaded in the ground plane, in *Antennas and Propagation 11th international Conference, April, 2001*, vol. 2, pp. 623–626
23. Shun-Yun Lin, Kin-Lu Wong, Effects of slotted and photonic bandgap ground planes on the charaetsristics of an air-substrate annular-ring patch antenna at TM21 mode. Proc. APMC **2**, 655–658 (2001)
24. L.-T. Wang, J.-S. Sun, The Compact, Broadband Microstnp Antenna with defective ground plane, in *IEE International Conference on antenna and Propagation*, vol. 2, April. 2003, pp. 622–624
25. Haiwen Liu, Zhengfan Li, Xiaowei Sun, Junfa Mao, Harmonic suppression with photonic bandgap and defected ground structure for a microstrip patch antenna. IEEE Microwave Wirel. Compon. Lett. **15**(2), 55–56 (2005)
26. N. Guerin, C. Hafner, X. Cui, et al., Compact directive antennas using frequency-selective surface (FSS), in *2005 Asia-Pacific Microwave Conference Proceedings*, vol. 1 (2005), p. 519
27. W. Richards, An improved theory for microstrip patches. IEE Proc. Part. H **132**, 93–98 (1985)
28. M. Ansarizadeh, A. Ghorbani, An approach to equivalent circuit modeling of rectangular microstrip antennas. Prog. Electromagnet. Res. B **8**, 77–86 (2008)
29. F. Abboud, J.P. Damiano, A. Papiernik, Simple model for the input impedance of coax-fed rectangular microstrip patch antenna for CAD, in *Microwaves, Antennas and Propagation, IEE Proceedings H*, vol. 135, no. 5 (1988), pp. 323–326
30. Jun-Seok Park, An equivalent circuit and modeling method for defected ground structure and its application to the design of microwave circuits. Microwave J. **46**(11), 22–38 (2003)
31. S. Hai, H. Guang, H. Wei, A broadband dual-polarized triangle patch antenna with DGS. J. Microwaves **21**(4), 27–30 (in Chinese) (2005)
32. https://www.3ds.com/products-services/simulia/products/cst-studio-suite/
33. T. Weiland, Time domain electromagnetic field computation with finite difference methods. Int. J. Numer. Model.: Electron. Networks, Devices Fields **9**(4), 259–319 (1998)
34. J. Geng, R. Jin, W. Wang, W. He, M. Ding, Q. Wu, X. Rui, G. Yang, Z. Fang, A new quasi-omnidirectional vertical polarisation antenna with low profile and high gain for DTV on vehicle. Microwaves, Antennas Propag., IET, Aug. **1**(4), 918–924 (2007)
35. J. Chen, M. Ding, J. Geng, X. Guo, R. Jin, Mattias lampe, Christoph Weiler, "Antenna Module", Chinese patent CN 2008000738, 08733941.2-1248 Europe PCT
36. J. Chen, R. Jin, J. Geng, Q. Wu, Z. Fang, Invention patent, "Dual band omnidirectional/directional antenna based on the EBG, patent number, ZL200610117044.8
37. J.P. Geng, J.J. Li, R.H. Jin, S. Ye, X.L. Liang, M.Z. Li, The development of the curved microstrip antenna with defected ground structure. Prog. Electromag. Res. PIER **98**, 53–73 (2009)

The manufacturer's authorised representative in the EU is Springer Nature Customer Service Centre GmbH, Europaplatz 3, 69115 Heidelberg, Germany. If you have any concerns regarding our products, please contact ProductSafety@springernature.com

Printed and bound by CPI Group (UK) Ltd, Croydon, CR0 4YY

27/07/2026

02177384-0001